Principles of Food Science and Nutrition

The Authors

Ms. S. Chandraprabha, completed MSc., in Food Science and Nutrition from Kerala Agricultural University. Presently working with SRM Faculty of Agricultural Science, SRM University, Tamil Nadu. She is author of 'Post Harvest Management and Value addition of Fruits and Vegetables' which is used as a textbook by B.Sc (Hons.) Agricultural Students. She has lot of experience in the area of Food Science, Post harvest Technology and having long teaching experience at Palar Agricultural College, Ambur.

Dr. K. Prakash, pursued Ph.D., in Fruit Science from Tamil Nadu Agricultural University, Presently Working in SRM College of Agricultural Science, SRM University, Tamil Nadu. He is expertise in Fruit Science and Post harvest Technology and having 6 years of teaching experience at JKK Munirajah College of agriculture, Erode and Palar Agricultural College, Ambur.

Mr. M. Balmurgan, pursuing doctoral degree in Food Science and Nutrition at Tamil Nadu Agricultural University. He has 3 years of Research experience in Post harvest Technology Centre, TNAU and 2 years of teaching experience at JKK Munirajah College of agriculture, Erode.

Dr. P. Sheela, completed her Ph.D., in Food Science and Nutrition from Tamil Nadu Agricultural University. She is currently working in SRM College of Agricultural Science, SRM University, Tamil Nadu.

Dr. A. Ravanachandar, pursued her Ph.D., in Species from TNAU. He has a several years teaching experience in JKK Munirajah College of agriculture, Erode. Currently Working in the SRM College of Agricultural Science, SRM University, Tamil Nadu.

Principles of Food Science and Nutrition

— *Authors* —

S. Chandraprabha

K. Prakash

M. Balmurugan

P. Sheela

A. Ravanachandar

2022

Daya Publishing House®

A Division of

Astral International Pvt. Ltd.

New Delhi – 110 002

ISBN: 978-93-5461-146-9 (HB)

Publisher's Note:

Every possible effort has been made to ensure that the information contained in this book is accurate at the time of going to press, and the publisher and author cannot accept responsibility for any errors or omissions, however caused. No responsibility for loss or damage occasioned to any person acting, or refraining from action, as a result of the material in this publication can be accepted by the editor, the publisher or the author. The Publisher is not associated with any product or vendor mentioned in the book. The contents of this work are intended to further general scientific research, understanding and discussion only. Readers should consult with a specialist where appropriate.

Every effort has been made to trace the owners of copyright material used in this book, if any. The author and the publisher will be grateful for any omission brought to their notice for acknowledgement in the future editions of the book.

Published by : **Daya Publishing House®**
A Division of
Astral International Pvt. Ltd.
– ISO 9001:2015 Certified Company –
4736/23, Ansari Road, Darya Ganj
New Delhi-110 002
Ph. 011-43549197, 23278134
E-mail: info@astralint.com
Website: www.astralint.com

Digitally Printed at : **Replika Press Pvt. Ltd.**

Preface

"Our health lies in the hand through the food we consume"

Food science is a field of science, studies the biochemical, physical, chemical nature of food, causes of deterioration, preservation and processing of foods. Nutrition is the science which interprets the nutrients and other substances present in food have a relationship to maintenance, growth, reproduction, health and disease of an organism. Food science and nutrition has gained added significance with an increase in the number of lifestyle related diseases, such as high blood pressure, atherosclerosis, heart diseases, diabetes mellitus, and obesity.

Food preservation is an action or a method of maintaining foods at a desired level of properties or nature for their maximum benefits. Processing and Preservation plays an important role to sustain agricultural production. Hereby, it helps to add value to fresh produce, thus having great scope for employment and income generation to the society.

This book covers a wide spectrum in the area of food science and nutrition which make it more nutritious and better for people. Thus helps in understanding the effects of each preservation method and handling procedure on foods is critical process in food processing. The book has been written keeping in view the requirements of the graduate and post-graduate students, teachers and research scientists in Food Technology as per the latest recommendations of the 5th Deans' Committee of the ICAR. The topics included in this text are of utmost significance and bound to provide incredible insights to readers.

S. Chandraprabha

K. Prakash

M. Balmurugan

P. Sheela

A. Ravanachandar

Contents

Chapter

1

Food Science: Definition, Scope, Classification and Food Pyramid

Food is the basic necessity of man. It is a mixture of different nutrients such as carbohydrate, protein, fat, vitamins and minerals. These nutrients are essential for growth, development and maintenance of good health throughout life. The selection of foods best suited for promoting health has been found out by trial and error by continued use. The knowledge of nutrition has been built upon the observation by several pioneers. Lavoisier is referred as father of nutrition due to his remarkable contribution in quantification of energy metabolism.

Define Food

Food can be defined as **anything edible that can be solid, semi-solid or liquid** which when **swallowed, digested and assimilated, nourishes the body**. These substances not only keep the person alive, but also **provide energy used for growth and development, regulate the body processes and protect the body from diseases.**

Definition – Food Science

Food science is the applied science devoted to the study of food. Food science is the study of the physical, biological, and chemical makeup of food; and the concepts underlying food processing.

Food Science has been defined by **Margaret (1968) as**: "The application of the basic **science and engineering to study the fundamental, physical, biological and biochemical nature of foods and the principles of processing and marketing of food".**

Dennis R Helmand, the International Food Technologists (IFT) president in October 2006 has described Food Science as an discipline in which engineering, biological and physical science are used to study the nature of foods, the cause of deterioration, the principles underlying food processing and improvement of foods for consuming public.

Define Food Technology

Food technology is the application of food science to the selection, preservation, processing, packaging, distribution, and use of safe food.

Food Technology can be defined as: **"science dealing with knowledge of doing things efficiently and effectively".**

Define Nutrition

Nutrition has been defined as that condition which **permits the development and maintenance of the highest state of fitness**. Fitness implies to good health, maximum capacity for work, the ability to undertake mental and physical tasks and the power to withstand physical, physiological and psychological stress.

Define Nutrients

- ✰ It can be defined as those **constituents in food that must be supplied to the body in sufficient amounts in order to grow, reproduce and lead a normal healthy life.**
- ✰ Various nutrients are protein, carbohydrates, fats, vitamins, minerals and water.

Food additive are non-nutritive substances added intentionally to food, generally in small quantities to improve its appearance, flavor, texture or storage properties. *e.g.* Preservatives, colourants, sweeteners *etc.*

Fermented foods are produced by the action of bacteria or moulds which act on carbohydrates and proteins present in foods and hydrolyse them to simpler products yielding predigested foods. *e.g.* idli, dosa, yoghurt.

Food chemistry is involved in investigating the composition, structure and properties of food stuffs and their components.

Food security is known as situation in which people do not live in hunger or fear of starvation.

Food fortification: The process whereby nutrients are added to foods in relatively small quantities to maintain or improve the quality of the diet of a group, a community or a population. *e.g.* iodized salt

Bio-fortification: The process of improving nutrient content in crops through breeding or genetic modification. *e.g.* golden rice

Functional Food

Foods that provide additional health-promoting and disease-preventing benefits beyond the basic function of supplying nutrients. A functional food may be a traditional food which we consume as part of our usual diets; it is shown to have physiological benefits and reduce the risk of chronic disease. They have positive effect on a person's health

Examples of Functional Foods

- Cereals: Wheat bran, oats, corn
- Legumes: Soybean and soya based products
- Oilseeds: Flaxseed, gardencress seed
- Vegetables: Carrots, green leafy vegetables, tomatoes, onion, garlic
- Fruits: Citrus fruit

Phytochemicals are non-nutrient compounds found in plant derived food that have biological activity in the body.

Food safety and regulation are related to food sanitation in public health and rules and regulations governing it.

Antioxidants is a compounds that protect biological systems against the potentially harmful effects of processes or reactions that can cause excessive oxidation

Scope of Food Science

Food Companies

Most packed food brands (Nestle, *etc.*) employ Food science graduates in various departments (Research, production, quality *etc.*).

Hospitals

Every good hospital employs a few clinical nutritionists and dieticians to check on patients.

Independent Consultation and Practice

After getting a few years of experience, you can become a consultant for health centres (Diabetic centre, fitness, weight loss *etc.*) depending on your area of expertise.

Research

You may get into industrial or academic research. Also, hotels, food safety, quality control, *etc.* and more.

Teachers: Schools, colleges and research institutions.

Scientist: in CFTRI, Mysore, DFRL, Hyderabad, NIN and research institutions *etc.*

Government organization: CDPO (Community Developmental Project Officer).

Classification of Foods

Man must eat to live and what he eats will affect to a high degree his ability to keep well, to work and to live long. Food performs many vital functions in the body. There are different food groups like cereals, pulses, fruits, vegetables, milk and meat products and nuts and oil seeds *etc.* Foods are classified based on their functions and nutritive value.

Classification of Foods

Functional Classification	*Nutritional Classification*
✰ Energy Yielding Foods	✰ Basic Four
✰ Body Building Foods	✰ Basic Five
✰ Protective and Regulatory Foods	✰ Basic Seven
✰ Health Maintaining Foods	✰ Basic Eleven

1. Functional Classification

In functional classification foods are classified based on their function in our body.

(a) Energy Yielding Foods

Foods rich in carbohydrates, fats and protein are called energy yielding foods. They provide energy to carry out various professional, household and recreational activities. Each one gram carbohydrate and protein gives 4 Kcal. While one gram of fat gives 9 Kcal.

Example: Cereals, roots and tubers, dried fruits, Nuts and oil seeds-fat and protein, Oils, butter, ghee-fat, Sugar.

(b) Body Building Foods

Foods rich in protein are called body building foods. Protein helps in building and repairing tissues. They are divided in to two groups.

- ✰ **Animal source:** Egg, milk, meat, and fish *etc.* - high biological value
- ✰ **Plant source:** Pulses, nuts and oil seeds - low biological value

(c) Protective and Regulatory Foods

Foods rich in protein, minerals and vitamins are known as protective and regulatory foods. They are essential for health and regulate activities such as maintenance of body temperature, muscle contraction, control of water balance, clotting of blood, removal of waste products. Protective foods are classified into two groups.

- Foods rich in vitamins, minerals and protein -milk, egg, fish, liver
- Foods rich in vitamins and minerals only - green leafy vegetables, vegetables and fruits

(d) Health Maintaining Foods

Foods rich in phytochemicals and antioxidants are called health maintaining foods. They help to prevent diseases like cancer, heart disease *etc.*

Example: whole grains, soy bean, green leafy vegetables, fruits and vegetables and spices *etc.*

2. Nutritional Classification

Foods have been classified into different groups depending upon the nutritive value, for the convenience of planning diets. Food groups like 'Basic four', 'Basic five', 'Basic seven' or 'Basic eleven' can be used for planning diets as per the convenience.

(a) Basic Four

Food Group	*Nutrients*
1. Cereals, millets and pulses	Energy, protein, B-vitamins, minerals
2. Vegetables and fruits	Vitamins, minerals and fibre
3. Milk, milk products, and animal foods	Protein, calcium, B-vitamins
4. Oils, fats, nuts and oilseeds	Energy, protein, fat and fat soluble vitamins

(b) Basic Five: ICMR Classification

Food Group	*Nutrients*
1. Cereals, Grains and its Products: Rice, Wheat, Ragi, Bajra, Maize, Jowar, Barley, Rice Flakes, Wheat Flour.	Energy, protein, invisible fat, thiamine, folic acid, riboflavin, iron and fibre
2. Pulses and legumes: Bengal gram, Black gram, Green gram, Red gram, Lentil (whole as well as dhals) Cowpea, Peas, Rajmah, Soyabeans, Beans.	Protein, invisible fat, thiamine, riboflavin, folic acid, calcium, iron and fibre

Food Group	*Nutrients*
3. **Milk and Meat products:** Milk, Curd, Skimmed milk, Cheese **Meat:** Chicken, Liver, Fish, Egg, Meat	Protein, fat, riboflavin, calcium, iron, vit 'A'
4. **Fruits and Vegetables:** **Fruits:** Mango, Guava, Tomato, Papaya, Orange. Sweet Lime, Watermelon **Vegetables (Green Leafy):** Amaranth, Spinach, Drumstick leaves, Coriander leaves, Mustard leaves, and fenugreek leaves. **Other Vegetables:** Carrots, Brinjal, Ladies fingers, Capsicum, Beans, Onion, Drumstick, Cauliflower.	Carotenoids, vitamin C, riboflavin, folic acid, iron, potassium, phosphorus and fibre.
5. **Fats and Sugars:** **Fats:** Butter, Ghee, Hydrogenated oils, Cooking oils like Groundnut, Mustard, Coconut. **Sugars:** Sugar, Jaggery	Energy, essential fatty acids, iron

(c) Basic Seven

Group	*Nutrients*
1. Green and yellow vegetables	Carotenoids, fibre
2. Oranges, grape fruit, tomatoes or raw cabbage	Vitamin C, carotenoids
3. Potatoes, other vegetables	Carbohydrate
4. Milk and milk products	Protein, fat, calcium, phosphorus
5. Meat, poultry, fish and eggs	Protein, fat, iron, vitamin A,
6. Bread, flour and cereals	Protein, thiamine, riboflavin, iron and fibre
7. Butter or fortified margarine	Fat and fat soluble vitamins.

(d) Basic Eleven

Group	*Nutrients*
1. Cereals	Carbohydrate (starch), thiamine, riboflavin, and fibre
2. Pulses	Protein, starch, fibre
3. Nuts and oil seeds	Fat, protein and minerals

Group	*Nutrients*
4. Vegetables	Vitamins, minerals and fibre
5. Fruits	Vitamins, minerals and fibre
6. Milk and milk products	Protein, fat, calcium, phosphorus
7. Eggs	Protein, fat, vitamin D, vitamin A
8. Flesh foods, fish	Protein, iron, calcium, phosphorus, vitamin B_6, vitamin B_{12}, ω-3 fatty acids
9. Fats and oils	Fat and fat soluble vitamins
10. Sugar and jaggery	Energy and iron
11. Spices and condiments	Phytochemicals and antioxidants

What is Food Pyramid?

A **food pyramid** or diet **pyramid** is a **pyramid**-shaped diagram representing the optimal number of servings to be eaten each day from each of the basic **food** groups.

The first food pyramid was published in **Sweden in 1974.** The food pyramid introduced by the United States Department of Agriculture in the year 1992 was called the "Food Guide Pyramid". It was updated in **2005** and then **replaced by My Plate in 2011.**

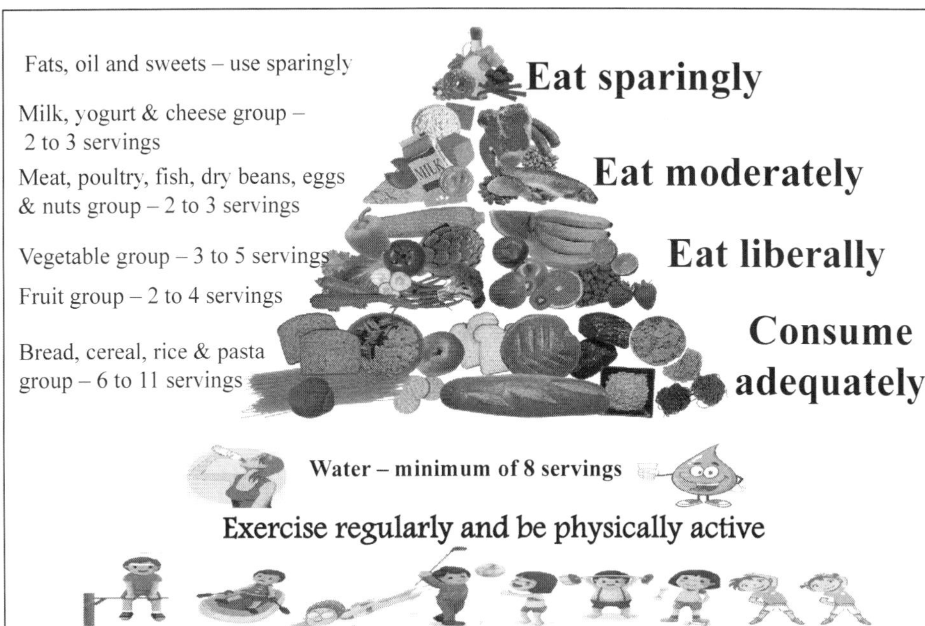

Food Pyramid.

My Pyramid (2015)

Food Groups	*Recommendation*	*Daily Amount (g or ml)*	
		Men	*Women*
Grains:	Half of grains consumed should be whole grains.		
Cereals and millets		360	330
Pulses		75	60
Vegetables:	Variety of vegetables should be taken		
Green leafy vegetables		100	100
Other vegetables		200	200
Roots and tubers		200	200
Fruits	Eat variety of fruits	100	100
Milk	Eat low - fat or fat free dairy products	300	300
Meat	Eat lean cuts, sea foods and avoid frying	100	100
Oil	Most oil should be from fish, nuts and vegetable oil Limit solid fat like butter, margarine or lard Limit the consumption of saturated fat, trans fat	20	20
Sugar	Use jaggery, avoid cane sugar	20	20

Chapter 2

Cooking Methods of Food

Cooking

The process of subjecting foods to the action of heat is termed as cooking. Heat may be transferred to the foods during cooking by conduction, convection and radiation or by radiation of microwaves.

Objectives/Merits of Cooking

- ☆ To sterilize the food
- ☆ To softens the food
- ☆ To improve palatability, appearance, flavour, texture and taste of food
- ☆ To increases the food consumption
- ☆ To increases availability of nutrients

Demerits of Cooking

- ☆ Heat sensitive nutrients (vitamins) may be lost during cooking. Water soluble vitamins are leached into the water during cooking.
- ☆ Vitamin A and C content may be reduced due to oxidation and heat.
- ☆ Quality of protein may be reduced due to destruction of certain amino acids during cooking *e.g.* bread

Classification of Cooking Methods

The methods of cooking are divided into groups because of the way food is cooked and the type of heat that is used.

Cooking Methods

Moist Heat (Water or steam)	***Dry Heat (Air or fat)***	***Combination***
Boiling	Roasting	Braising
Simmering	Grilling	
Poaching	Toasting	
Stewing	Baking	
Blanching	Satueing	
Steaming	Frying	
Pressure cooking		

Moist Heat Cooking Methods

In moist heat cookery methods, liquid is used as a medium to cook the food. Such medium could be water, coconut cream or oil. These liquids are added to the food before heat is applied to it or sometimes heat is applied to the liquid before the food is added into the cooking utensils to be cooked.

Boiling

Boiling is a method of cooking foods by just immersing them in water at 100°C and maintaining the water at that temperature till the food is tender. Rice, egg, dhal, meat, roots and tubers are cooked by boiling.

Merits

1. Simple method- It does not require special skill and equipment.
2. Uniform cooking can be achieved.
3. Protein gets denatured, starch gets gelatinized and collagen gets hydrolysed

Demerits

1. Continuous excessive boiling leads to damage in the structure and texture of food.
2. Loss of heat labile nutrients such as B and C vitamins if the water is discarded.
3. Time consuming– Boiling takes more time to cook food and fuel may be wasted.
4. Loss of colour– water soluble pigments may be lost.

Simmering

When foods are cooked in a pan with a well-fitting lid at temperature just below the boiling point 82-99°C of the liquid in which they are immersed the process is known as simmering. It is a useful method when foods have to be cooked for a long time to make it tender as in case of cheaper cuts of meat, fish, cooking custards, kheer, vegetables and carrot halwa. This method is also employed in making soups and stocks.

Merits

1. Foods get cooked thoroughly
2. Scorching or burning is prevented
3. Losses due to leaching is minimum

Demerits

1. There is loss of heat sensitive nutrients, due to long period of cooking
2. Takes more time and more fuel

Poaching

This involves cooking in the minimum amount of liquid at temperatures of 80-85°C that is below the boiling point. Egg and fish can be poached.

Merits

1. No special equipment is needed.
2. Quick method of cooking and therefore saves fuel.
3. Poached foods arc easily digested since no fat is added.

Demerits

1. Poached foods may not appeal to everybody as they are bland in taste.
2. Food can be scorched if water evaporates due to careless monitoring.
3. Water soluble nutrients may be leached into the water.

Stewing

It refers to the simmering of food in a pan with a tight fitting lid using small quantities of liquid to cover only half the food. This is a slow method of cooking. The liquid is brought to boiling point and the heat is reduced to maintain simmering temperatures (82-90°C).The food above the liquid is cooked by the steam generated within the pan. Apple, meat along with roots, vegetables and legumes are usually stewed.

Merits

1. Loss of nutrients is avoided as water used for cooking is not discarded.
2. Flavour is retained.

Demerits

1. The process is time consuming and there is wastage of fuel.

Blanching

In meal preparation, it is often necessary only to peel off the skin of fruits and vegetables without making them tender. This can be achieved by blanching. In this method, food is dipped in boiling water for 5 seconds to 2 minutes depending on the texture of the food. This helps to remove the skin or peel without softening food. Blanching can also be done by pouring enough boiling water on the food to immerse it for some time or subjecting foods to boiling temperatures for short periods and then immediately immersing in cold water. The process causes the skin to become loose and can be peeled off easily.

Merits

1. Peels can easily be removed to improve digestibility.
2. Destroys enzymes that bring about spoilage.
3. Texture can be maintained while improving the colour and flavour of food.

Demerits

1. Loss of nutrients if cooking water is discarded.

Steaming

It is a method of cooking food in steam generated from vigorously boiling water in a pan. *e.g*. Idli, custard, idiappam and vegetables

Types of Steaming

- ✰ **Wet steaming-** here steam is direct contact with the food
- ✰ **Dry steaming -** here food is cooked in double boiler and food does not have any direct contact with the food.

Merits

1. Less chance of burning and scorching.
2. Texture of food is better as it becomes light and fluffy. *e.g.* Idli.
3. Cooking time is less and fuel wastage is less.
4. Steamed foods like idli and idiappam contain less fat and are easily digested and are good for children, aged and for therapeutic diets.
5. Nutrient loss is minimised.

Demerits

1. Steaming equipment is required.
2. This method is limited to the preparation of selected foods.

Pressure Cooking

In pressure cooking, escaping steam is trapped and kept under high pressure. So the temperature of boiling water can be raised above 100° C. So the foods get cooked quickly. Boiling for 1 hour at 100° C can be reduced to half an hour at 110° C and further can be reduced to 1/4 hour at 120° C. The domestic pressure cooker is usually made of heavy gauge aluminium and the pressure is regulated by weights placed upon a release valve. A safety valve is fitted to prevent building up beyond the safe level.

Example: Rice, dhal, meat, roots and tubers are usually pressure cooked.

Merits

1. It takes less time to cook
2. Requires less attention
3. Nutrient and flavour loss is minimized
4. Conserves fuel and time as different items can be cooked at the same time
5. Less chance for burning and scorching

Demerits

1. Thorough knowledge of using the equipment is required. Otherwise accidents can happen
2. Careful watch on the cooking time is required to prevent over cooking.
3. The initial investment may not be affordable to everybody

Dry Heat Cooking Methods

In dry heat cooking methods, the food being cooked does not use water to cook the food. The food is left dry and heat is applied to cook the food. In this method air or fat/oil is used as cooking medium.

Roasting

In this method food is cooked in a heated metal or frying pan without covering it. In roasting, direct heat is applied to the food evenly on all sides with temperatures of at least 150 °C.

Merits

1. Improves the colour, flavor and texture of foods
2. Reduces the moisture content of the food and improves the keeping quality. *e.g.* Rava
3. It is easy to powder *e.g.* Cumin seeds and coriander seeds after roasting
4. It is one of the quick methods of cooking foods

Demerits

1. Constant attention is required
2. Losses of nutrients like amino acids occur when the food becomes brown

Grilling or Broiling

Grilling consists of placing the food below or above or in between a red hot surface. When under the heater, the food is heated by radiation only. This results in browning of foods. Then the heat is more slowly conducted through the surfaces of the food downward. As heating is most superficial, grilled foods are usually reversed or rotated. If the food is above the heater, heat is transmitted to the food through convection current as well as radiations with consequent increased efficiency. Foods cooked by grilling are cob on the corn, papad, brinjal, phulkas, sweet potato. Barbecues are also made by this method

Merits

1. Enhances flavour, appearance and taste of the product.
2. It requires less time to cook.
3. Minimum fat is used.

Demerits

1 Constant attention is required to prevent charring.

Toasting

This is a method where food is kept between two heated elements to facilitate browning on both sides. This can be adjusted to give the required degree of brownness through temperature control. Bread slices are cooked by toasting.

Merits

1. Easy and quick method.
2. Flavour improved.

Demerits

1. Special equipment required.
2. Careful monitoring is needed to prevent charring.

Baking

In this method, the food gets cooked in an oven by dry heat. The temperature range maintained in an oven is 120°C-260°C. The food is usually kept uncovered in a container greased with a fat coated paper. In this method heat is transferred by radiation, convection and conduction. Example: Bread, cake, biscuits, pastries and meat are prepared by this method.

Merits

1. Foods become light and fluffy, *e.g.*, cakes, custards, bread.
2. Uniform and bulk cooking can be achieved, *e.g.*, Bun, bread.
3. Flavour and texture are improved.
4. Variety of dishes can be made.

Demerits

1. Special equipment like oven is required.
2. Baking skills are necessary.
3. Careful monitoring needed to prevent scorching.

Sauteing

Sauteing is a method in which food is cooked in little oil just enough to cover the base of the pan. The food is turned occasionally. Foods cooked by sauteing are generally vegetables which are used as side dishes in a menu. In this method heat is transferred to the food mainly by conduction. Example: Dosa, vegetables

Merits

1. Takes less time.
2. Simple technique.
3. Minimum oil is used.

Demerits

1. Constant attention is needed as there is chance of scorching or burning.

Deep Fat Frying

In this method, the food is cooked by totally immersed in hot oil. The food is cooked at the temperature of 180-220° C. Generally 10 per cent of oil is absorbed but larger amount of oil is absorbed when oil is repeatedly used. Convection is the mode of heat transfer involved in this method.

Example: Samosa, chips, pakoda are examples of deep fat fried foods.

Shallow Fat Frying

Food is cooked in little amount of fat but not enough to cover it. In this method food is not totally immersed in oil. Heat is transferred to the food partially by conduction and convection.

Example: Omlette, cutlets, parathas.

Merits

1. Very quick method of cooking.
2. The calorific value of food is increased since fat is used.
3. Frying gives good flavour and attractive appearance.
4. Taste and texture are improved.

Demerits

1. Careful monitoring is required.
2. The food may become soggy.
3. Fried foods are not easily digested.

Combination of Cooking Methods

Braising

Braising is a combined method of roasting and stewing in a pan with a tight fitting lid.

Examples

Uppuma- Roasting and boiling

Meat cutlet - Boiling and deep frying.

Vermicelli payasam - Roasting and simmering

Mutter paneer - Frying and stewing

Vegetable curry - Sauteing and simmering

Vegetable pulav - Frying and stewing

Microwave Cooking

In Microwaves cooking electromagnetic waves are used to cook the food. Magnetron is the power source of electromagnetic waves. When the electromagnetic waves are interacted with the water dipole present in the food, the molecules begins to vibrate (2,450 million times/sec.). The vibration produces friction that produce heat and food get cooked.

Magnetron

↓

Electromagnetic waves

↓

Absorbed by food

↓

Interact with water dipole

↓

Vibration of molecules (2450 times/seconds)

↓

Creation of Friction

↓

Production of Heat

↓

Cooking

Merits

1. Quick method.
2. Food gets cooked uniformly.
3. Flavour and texture do not change when reheated in Microwave oven.
4. Nutrient loss is minimised.
5. Preserve the natural colour of the fruits and vegetables.
6. No fat or low fat cooking can be made
7. Washing dishes is much easier as food does not stick with the vessels.

Demerits

1. Baked products do not get a brown surface.
2. Microwave cooking cannot be used for simmering, stewing or deep frying.
3. Special skill is required.

Solar Cooking

Solar cooker consists of a well insulated box which is painted black on the inside and covered with one or more transparent covers. The purpose of these transparent covers is to trap heat inside the solar cooker. These covers allow the radiation from the sun to come inside the box but do not allow the heat from the hot black absorbing plate to come out of the box. Because of this, temperature up to 140°C can be obtained which is adequate for cooking.

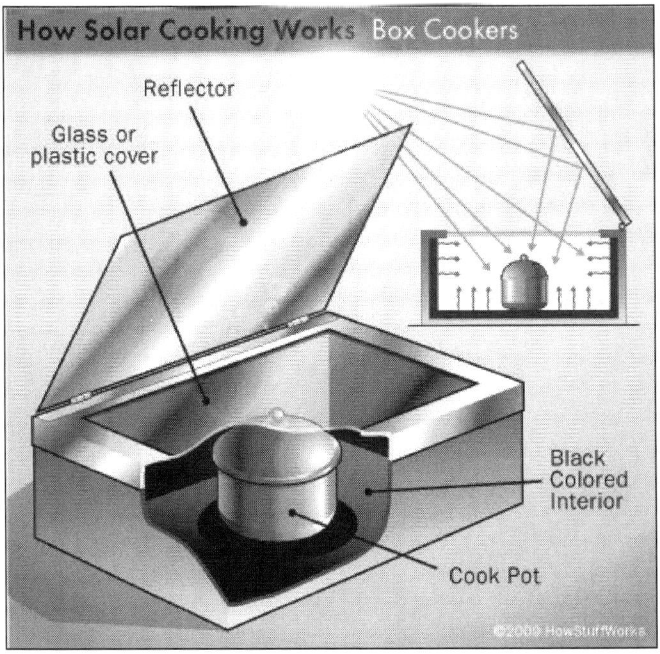

Important Parts of Box Type Solar Cooker

- ☆ **Outer box**: made of wood, iron sheet or fiber
- ☆ **Inner box**: made of galvanised iron or aluminium sheet and coated with black paint
- ☆ **Mirror**: to increase the solar radiation input
- ☆ **Cooking containers**: made of aluminium or stainless steel and painted dull black on the outer surface
- ☆ **Thermal insulation**: to prevent heat losses from the cooker
- ☆ **Double glass covers**: to trap heat inside the solar cooker

Merits

1. Simple technique.
2. Cost effective as sunlight is the source of energy.
3. Original flavour of food is retained.
4. There is no danger of scorching or burning.
5. Loss of nutrients is minimized.

Demerits

1. Special equipment is needed.
2. Slow cooking process.
3. Cannot be used in the absence of sunlight, rainy season, late evening and night.

Chapter 3

Scope and Importance of Nutrition and the Relation of Nutrition to Health

Introduction

Food is comprised of nutrients that are classified by their rolc in the body. the **energy-yielding macro nutrients (carbohydrates, protein and fat), the essential micro nutrients (vitamins, minerals, and water), and numerous other components.** Although micro nutrients do not supply energy to fuel the body, they are indispensable for the proper functioning of the metabolic and regulatory activities in the body. Other non-essential nutrients, such as flavonoids, phytoestrogens, carotenoids and probiotics, also may have important health-promoting properties.

The daily intake of a variety of foods provides energy and nutrients that are essential to the health and well-being of an individual. The relationships among food intake, nutrition, and health define the field of nutrition. More fully, **nutrition is the study of food, its nutrients and chemical components, and how these constituents act and interact within the body to affect health and disease.**

Importance of Nutrition

- ☆ Eating a balanced diet is vital for good health and wellbeing. Food provides our bodies with the energy, protein, essential fats, vitamins and minerals to live, grow and function properly. We need a wide variety of different foods to provide the right amounts of nutrients for good health.
- ☆ Nutrition, nourishment, or aliment, is the supply of materials from food and it's required by organisms and cells to stay alive. In science and human

medicine, nutrition is the science or practice of consuming and utilizing foods.

- In hospitals, nutrition may refer to the food requirements of patients, including nutritional solutions delivered via an IV (intravenous) or IG (intragastric) tube.
- Nutritional science studies how the body breaks food down (catabolism) and repairs and creates cells and tissue (anabolism) - catabolism and anabolism = metabolism. Nutritional science also examines how the body responds to food.
- Nutrition also focuses on how diseases, conditions and problems can be prevented or lessened with a healthy diet.
- In addition, nutrition involves identifying how certain diseases, conditions or problems may be caused by dietary factors, such as poor diet (malnutrition), food allergies, metabolic diseases, *etc.*,

Scope of Nutrition

Food Services: This can include a number of facilities in the commercial sector like food manufacture, catering services and restaurants where nutrition professionals can do anything from menu planning to meal preparation to promotion of the food products.

Health Care: This is one of the largest and well-known functions of nutrition in hospitals and clinics. Dieticians assist in treating patients with some big hospitals also providing scope for research, food administration, teaching *etc.*

Dissemination: This entails producing books, articles, promotions, television programmes on optimum dietary practices, since the present era is highly health-conscious.

Institutional Catering: Nutrition and dietetics professionals are needed to plan and prepare nutritious and well-balanced meals for schools, colleges, factories, offices, canteens *etc.*

Research and Development: R and D deals with conducting research projects on various food items to ensure welfare from both the commercial food services viewpoint, plus that of the health care provision.

Social welfare: Run by governmental organizations, this section is busy in improving the eating habits and consequently, the health of the less-fortunate groups in society.

Relation of Nutrition to Health

Health is related to the food consumed. To maintain good health, ingesting a diet containing the nutrient in correct amount is essential. A balanced diet is one which contains different types of food in such quantities and proportions so that the need for calories, protein, fats, minerals and vitamins is adequately met and a small provision is made for extra nutrients to withstand short duration of leanness. Deficiency of any nutrient affects health of an individual. Food has not only nutrients but also nutraceuticals which prevent degenerative diseases.

- Nutrition is a process in which the food that is consumed is used for **nourishing the body.** Good nutrition is essential for good health.
- Nutrition is very important for a **person to grow and develop normally and to remain healthy throughout life.**
- When a person does not eat proper food, there are chances of his not developing normally- the likelihood that some **organ of his body may start malfunctioning**, or that he may get some disease
- Poor nutrition may also influence his mental and social well-being.
- Thus, generally, the health of a person depends on the kind and amount of food he eats.
- The right kind of food eaten in the **right quantity is necessary for good health.**

Health -is defined by the World Health Organisation as the 'State of complete physical, mental and social well-being and not merely the absence of disease and infirmity'.

Diet - It includes the kinds and amount of food and drinks consumed.

Therapeutic diet - The diet changed according to the physical, mental and social well being of a person.

Over Nutrition - Taking excess of one or more nutrients. *e.g.* Obesity

Under Nutrition - Deficiency of one or more nutrients. *e.g.* Anemia, Pellagra, PCM (protein calorie mal nutrition) goiter, night blindness.

Mal nutrition - Either excess or deficiency of one or more nutrients

Food in Relation to Health

Nutrient	*Source*	*Deficiency*
Energy, Protein	Fat, sugar, cereals, pluses, egg, meat	Underweight, marasmus, kwashiorkor
Calcium	Milk, green leafy vegetables, gingelly seed	Rickets, osteomalacia, tetany
Iron	Liver, green leafy vegetables, rice flakes, jaggery.	Anaemia
Vitamin A	Liver, egg yolk, butter, green leafy vegetables, carrots	Night blindness
Thiamine	Yeast, outer layers of cereals, pulses, nuts.	Pain in the calf muscle, weakness of heart muscle

Nutrient	*Source*	*Deficiency*
Niacin	Groundnuts, whole cereals, pulses	Dementia, diarrhea, dermatitis
Vitamin B_6	Meat, liver, vegetables, whole cereal grains	Anaemia, angular stomatitis
Folic acid	Fresh green leafy vegetables, lady's finger, cluster beans	Megaloblastic anaemia
Vitamin B_{12}	Yeast, fermented foods	Pernicious anaemia
Vitamin C	Citrus fruit, amla, guava	Bleeding gums
Vitamin D	Eggs, fresh foods, sunlight	Rickets, osteomalacia

Balanced Diet for Energy

Daily allowances of essential vitamins, minerals, fiber, protein and other nutrients assist in cellular growth and the many body processes that keep you alive and functioning well. A balanced diet lets you consume these nutrients within reasonable calorie limits. Eating a variety of foods from all the food groups provides plenty of energy without overloading on high-calorie proteins, fats and sugar.

Nutrition for Strength

Many of these same foods help to establish the type of strength you will need to maintain an active lifestyle. Along with milk products, fish provides calcium, which is needed for musculoskeletal strength. The National Osteoporosis Foundation relates the importance of adequate calcium stores for healthy bones. Muscle activity also requires calcium, while antioxidant vitamins A, C and E, found in many fruits and vegetables, support a strong immune system. B vitamins, iron, potassium and other minerals encourage cardiovascular strength.

Nutrition for Stress Relief

Physical and mental tension causes strain on many systems of the body, such as the cardiovascular and immune systems. Therefore, optimum health depends upon stress management, in which nutrition plays a vital role. The Centres for Disease Control and Prevention note that a healthy diet prepares you to handle daily stress and to take part in an exercise routine, which also controls stress levels.

Diet and Exercise for Longevity

Individuals who practice stress relief can live longer, more active lives than those who resort to unhealthy coping habits, such as eating junk foods and drinking too much alcohol. Simply controlling your weight with a good diet and **regular exercise reduces your risk for Type 2 diabetes, heart attack and some types of cancer.** General In addition, getting adequate daily nutrients, such as fiber and calcium, can help to stave off chronic degenerative diseases, such as coronary heart disease and osteoporosis.

Nutritional status is the condition of health as it is related to the use of food by the body. Dietitian's physicians can observe physical signs that suggest good or poor nutrition as listed in the table

Body Parts	***Good Nutrition***	***Bad Nutrition***
Body size	Normal weight for height body frame and age, normal rate of growth	Recessive obesity or thinness, sudden loss or gain in weight failure to grow in stature to gain weight.
Behaviour	Alert, good attention span, regular attendance, cooperative, cheerful, interested, has endurance	Apathetic, short attention span, frequent absence. Irritable, easily fatigue, inability to concentrate, poor work capacity
Skin	Firm, clear, slightly moist, healthy, pink mucous membrane, firm pink nails	Dry, pale scaly around ears and nose, bed sores, brittle ridged nails
Hair	Soft, glossy, healthy scalp	Dry, brittle then easily pulled out, change in pigmentation
Eyes	Clear bright not unduly sensitive to light	Red, swollen or dry, itching or burning. Poor vision in dim light, excessive, sensitivity to bright light.
Mouth	Moist smooth, pink tongue with surface papillae, pink firm gums. Even teeth, well formed jaw.	Chopped tissues at corners of lips. Swollen tongue – scarlet or magneta smooth in appearance. Bleeding gums, receding from tooth line. Decayed or missing teeth. Inability to chew
Skeleton	Erect posture, arms and legs straight abdomen in, chest chin in	Poor posture, deformities of long bones, spines, pelvis, hump back, bow legs or knock ness
Neuro-muscular system	Firm, strong muscles with moderate padding of fat good muscular co ordination.	Flabby under developed muscles, lack of fat padding or excessive fat; poor muscle coordination reduced knee and ankle reflexes, burning and tingling of hands and feet.

Body Parts	*Good Nutrition*	*Bad Nutrition*
Gastro intestinal system	Good appetite and digestion, regular elimination	Poor appetite, in digestion, diarrhea or constipation
Glands	No enlargement of thyroid	Enlarged thyroid
Immune system	Resistance to infections	Frequent colds and other infections, longer convalence from illness slower wound healing

Malnutrition Across the Life Span

Malnutrition affects all age groups across the entire life span

Life Stage	*Common Nutritional Disorders*	*Main Consequences*
Embryo/fetus	Intrauterine growth retardation, Low birth weight, Iodine deficiency disorders (IDD), Brain damage	Folate deficiency, Neural tube defects, Stillbirths
Neonate	Low birth weight, Growth retardation, IDD	Developmental retardation, Brain damage, Early anaemia
Infant and young child	Protein-energy malnutrition (PEM), IDD, Vitamin A deficiency (VAD), Iron deficiency and anaemia (ID and A)	Continuing malnutrition, Developmental retardation, Increased risk of infection, High risk of death, Goitre, BlindnessAnaemia
Adolescent	PEM, IDD, ID and A, Folate deficiency, Calcium deficiency	Delayed growth spurt, Stunted height, Delayed/retarded intellectual development, Goitre increased risk of infection, Blindness, Anaemia, Inadequate bone mineralization
Pregnant and lactating women	PEM, IDD, VAD, ID and A,Folate deficiency, Calcium deficiency	Insufficient weight gain in pregnancy, Maternal anaemia, Maternal mortality, Increased risk of infection, Night blindness Low birth weight/ high-risk death rate for fetus

Life Stage	***Common Nutritional Disorders***	***Main Consequences***
Adults	PEM, ID and A,Obesity, Diet-related diseases	Thinness, Lethargy, Obesity, Heart disease, Diabetes, Cancer, Hypertension/stroke Anaemia
Older persons Diet-related diseases	PEM, ID and A, Osteoporosis, Obesity accidents, Diet-related diseases	Obesity, Spine/hip fractures, Heart disease, Diabetes and Cancer

Chapter 4

Carbohydrate: Classification, Function, Digestion, Absorption, Deficiency Symptoms, Sources and Requirements

Carbohydrate is one of the major nutrients. The term carbohydrate is derived from a French word which means "hydrate the carbon". Carbohydrates are organic compounds composed of carbon, hydrogen and oxygen, in the ratio of 1:2:1. The empirical formula of carbohydrate is $C_nH_{2n}O_n$. Carbohydrates are widely distributed in plants in the form of sugars, starches, celluloses and other complex substances. They are formed from carbon dioxide of the atmosphere by photosynthesis.

Classification of Carbohydrate

Carbohydrates are classified into four major groups. The classification is based on the number of sugar units they contain.

- ☆ Monosaccharide (simple sugars)
- ☆ Disaccharide
- ☆ Oligosaccharide
- ☆ Polysaccharide

Monosaccharide's (Simple Sugars)

They are often called as "simple sugars" which cannot be hydrolysed in to simple form. The general formula is $(CH_2O)n$. Monosaccharide's are soluble in water and sweet in taste. All Monosaccharide's are reducing in nature. Normal blood glucose level is 80 to 100 mg/100 ml of blood. If the blood glucose level reaches above 180 mg/100 ml (renal threshold level) glucose is excreted in urine.

Monosaccharides with Examples

Monosaccharides	*Examples*
Biose ($C_2H_4O_2$)	Glycolic Aldehydes
Triose ($C_3H_6O_3$)	Glyceraldehyde, Dihydroxy Acetone
Tetrose ($C_4H_8O_4$)	Erythrose, Threose
Pentose ($C_5H_{10}O_5$)	Arabinose, Xylose, Ribose, Deoxyribose
Hexose ($C_6H_{12}O_6$)	Glucose, Fructose, Galactose, Mannose

Glucose (Dextrose)

Glucose is otherwise known as dextrose/aldose sugar/grape sugar/corn sugar. Glucose serves as the main source of energy in the body. Normal human blood contains about 80-100 mg of glucose per 100ml of blood. This level is maintained constantly in healthy subjects.

Fructose (Levulose/Fruit sugar)

Fructose occurs in the Free State along with glucose in many fruits (2-5 per cent) and in honey (30-40 per cent). It is also utilized by the body as a source of energy.

Galactose

It does not occur in the free state but occurs as a constitute of lactose present in milk.

Disaccharide

They yield two molecules of the same or of different monosaccharide on hydrolysis. The general formula is $Cn(H_2O)_{n-1}$.

Example: Sucrose, Lactose, Maltose

$$\text{Maltose} + H_2O \xrightarrow{\text{Maltase}} \text{Glucose} + \text{Glucose}$$

$$\text{Sucrose} + H_2O \xrightarrow{\text{Sucrase}} \text{Glucose} + \text{Fructose}$$

$$\text{Lactose} + H_2O \xrightarrow{\text{Lactase}} \text{Glucose} + \text{Galactose}$$

i) Sucrose (Cane sugar/beet sugar)

Sucrose occurs in sugar cane (10-12 per cent) and beet root (12-18 per cent). It is manufactured on a large scale from sugar cane or beet root. Sucrose is hydrolysed to glucose and fructose by dilute mineral acids or by the enzyme sucrase present in intestinal juice.

ii) Maltose (Malt sugar)

Maltose is present in malt. It is formed in cereals grains (barley, jowar, ragi *etc.*) during germination by the hydrolysis of starch by the enzyme diastase (Amylase)

iii) Lactose (Milk sugar)

Lactose occurs in the milk of all mammals but has not so far been found in plant foods.cow and buffalo milk contains 4 per cent of lactose whereas human milk contains 7 per cent of lactose. Lactose is hydrolysed to glucose and galactose by the enzyme lactase present in the intestinal juice.

Oligosaccharides

Oligosaccharides contain 2-10 monosaccharides joined by glycosidic bonds. Oligosaccharides can be hydrolysed in to monosaccharides. Oligosaccharides present in pulses causes flatulence in human beings.

Types

1. Trisaccharide
2. Tetrasaccharide

Example: Raffinose, Stachyose

Polysaccharide

Polysaccharide contains many (more than 10 molecules) monosaccharide joined by glycosidic bonds. Starch is the main storage polysaccharide of plants. The general formula is $(C_6H_{10}O_5)X$. They are insoluble in water and tasteless in nature.

Types

1. Homopolysaccharide- Starch, Glycogen, Dextrin, Cellulose
2. Heteropolysccharide – Hemicellulose, gums

Starch

Starch is a polysaccharide formed in nature by the condensation of a large number (4000-15000) of glucose molecules. It consists of two components called amylose and amylopectin, Amylose is a linear chain polysaccharide with 500-5000 glucose molecules linked by α-1, 4 glycosidic linkage. Amylopectin is a highly branched chain polysaccharide with 50,000-500000 glucose molecules linked by α-1, 4 1,6 glycosidic linkage.

Dextrins

Dextrins are polysaccharides formed by the partial hydrolysis of starch by acids or amylase.

Glycogen

Glycogen is the reserve carbohydrate found in liver (3-7 per cent) and muscle (0.5-1 per cent) of animals and man

Carbohydrates

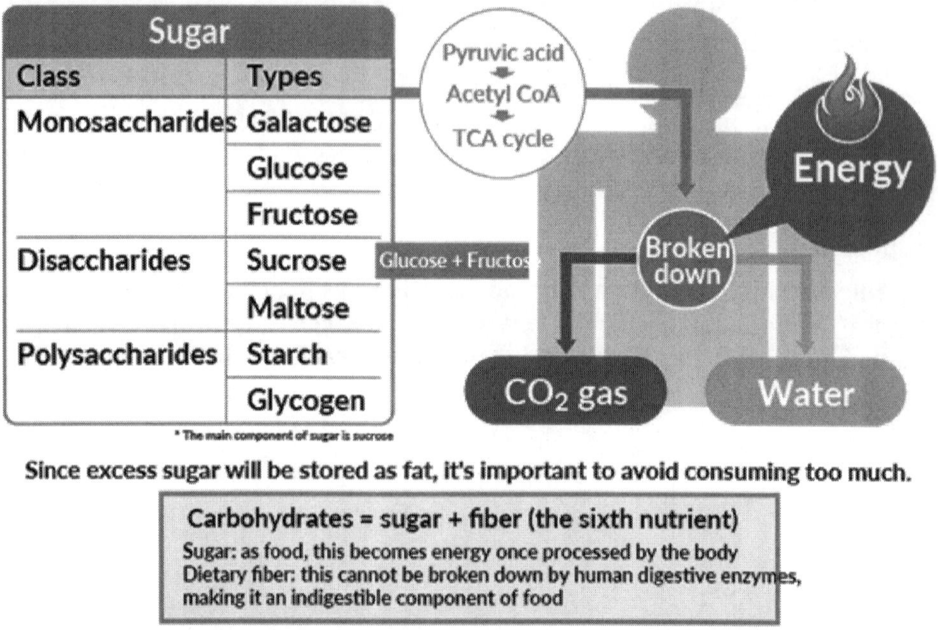

Functions of Carbohydrate

Carbohydrates perform the following functions.

1. Source of Energy

The carbohydrates are major source of energy for the body. Each gram of carbohydrate yields 4 Kcal of energy. In Indian diets 60 – 80 per cent of energy is derived from carbohydrate.

2. Protein Sparing Action

Carbohydrates exert a protein sparing action. If sufficient amounts of carbohydrates are not available in the diet, the body will convert protein to glucose in order to supply energy. Hence to spare proteins for tissue building, carbohydrates must be supplied in optimum amounts in the diet. This is called the protein sparing action of carbohydrates.

3. Fat Metabolism

Carbohydrates are essential to maintain normal fat metabolism. An insufficient amount carbohydrate in the diet results in larger amounts of fat being used for energy. This leads to accumulation of ketone bodies and produce ketosis.

4. Synthesis of Body Substances

Carbohydrates help in the synthesis of non-essential aminoacids, glycoproteins and glycolipids.

5. Promote the Growth of Beneficial Bacteria

Lactose encourages growth of beneficial bacteria in intestine.

6. Precursors of Nucleic Acid

Carbohydrates and products derived from them, serve as precursors of compounds like nucleic acids, connective tissue matrix and galactosides of nervous tissue.

7. Detoxification Function

Glucuronic acid, a metabolite of glucose and glycogen serves as a detoxifying agent. It protects liver from some harmful substances like carbon-tetrachloride, alcohol and arsenic.

8 Prevent Constipation

Insoluble fibres are carbohydrate which gives bulkiness to the intestinal contents and helps in elimination of wastes and prevent constipation.

9 Synthesis of Ribose

The pentose ribose is present in RNA and nucleotides. It is formed from glucose by the pentose path way.

Digestion of Carbohydrates

Digestion is the process of conversion of complex molecule into simpler molecules

Mouth

The first stage of digestion of carbohydrate takes place in the mouth. Saliva contains ptyalin or α- amylase enzyme. It converts starch to maltose.

Stomach

In stomach food gets mixed with gastric juice and some of the sucrose present in the food hydrolysed to monosaccharides by the action of HCl in the stomach.

Small Intestine

Food moves through intestine and react with pancreatic amylase. Small intestine contains duodenum, jejum and ileum. In duodenum, the entire starch molecules converted into disaccharides. Jejum and Ileum contain disaccharide enzyme like maltase, sucrose and lactase digested into monosaccharides with the help of intestinal mucosa.

$$\text{Maltose} + H_2O \xrightarrow{\text{Maltase}} \text{Glucose} + \text{Glucose}$$

$$\text{Sucrose} + H_2O \xrightarrow{\text{Sucrase}} \text{Glucose} + \text{Fructose}$$

$$\text{Lactose} + H_2O \xrightarrow{\text{Lactase}} \text{Glucose} + \text{Galactose}$$

Cellulose and other polysaccharides are not digested by enzymes. The undigested material present in the food are not acted upon the digestive juice. They add bulk to the content of large intestine is excreted in the faeces. Some of these are fermented by bacteria present in large intestine and eliminated.

Absorption of Carbohydrate

- The end products of carbohydrate digestion are monosaccharides – glucose, galactose and fructose.
- The monosaccharides are absorbed in to the mucosa cells of small intestine and pass into portal circulation. The glucose and galactose are absorbed by process of active transport. But the fructose is absorbed by diffusion.
- Then they are transported to liver. Here glucose, fructose and galactose are converted into glycogen for storage. The glycogen is reconverted into glucose as needed.

Digestion and Absorption of Carbohydrate

Starch/Glycogen/Disaccharides

↓ Digestion

Glucose/galactose/fructose

↓ Absorption

Mucosa cells of small intestine

↓

Portal circulation

↓

Liver

↓

Stored as glycogen or fat

Deficiency Symptoms

Acidosis: In carbohydrate starvation, there is shift from glycolysis (breakdown of glucose) to lipolysis (breakdown of lipids) and ketogenesis for energy needs. The resultant production of ketoacids increases acidity in the blood and other body tissues.

Ketosis: During prolonged carbohydrate fasting or starvation, acetyl-CoA in the liver is used to produce ketone bodies. It is known as accumulation of ketone bodies.

Hypoglycemia: The non availability of glucose due to severe lack of carbohydrate causes drop in the blood sugar levels. Hypoglycemia occurs when blood glucose, levels drop under 70 mg/dL with typical symptoms like giddiness, fatigue and distress

Dehydration and reduced body secretions: As there is loss of fluids from the body due to ketosis, a state of chronic dehydration is reached.

Loss of sodium: Excess of ketone in the blood leads to fluid loss and excretion of sodium ions (Na^+) from the body. This may lead to muscle cramps and tiredness

Fatigue and decreased energy levels: The immediate non availability of glucose in the blood for energy production, result in dip in the energy levels and fatigue.

Muscle wasting: As the fat reserves and amino acids are getting used up for energy production, there will be general loss of muscle mass and impairment of growth.

Unhealthy weight loss: The loss of fat and muscle mass leads to emaciation and weight loss. **Weakened immune system**: With increased fluid loss and degradation, vitamin C loss from the body is increased. Adding to this, the chronically dehydrated condition leads to weakened immune system and susceptibility to infections.

Constipation: Dietary fiber is an essential component of carbohydrate food, which is known to prevent recto-colon cancer and help digestion. The absence of dietary fiber can cause constipation.

Mood swing: Dietary carbohydrate exclusion causes the brain to stop regulating serotonin hormone. Low serotonin supply causes mood swing and depression.

Sources

The major sources are sugar, jaggery, sago, rice, dried dates, skimmed milk powder, whole milk powder, wheat, potatoes, sweet potatoes and sweets.

Many foods contain carbohydrates, including sweets, breads, cereals, dairy products, vegetables and soft drinks. Choose whole grains over refined grain products, and eat more fruits, vegetables and legumes than sugary and salty snacks, for enhanced nutrient and fiber intake.

Recommended Dietary Allowances (RDA)

Carbohydrate need to consume130g/day for adults and children. An average adult should consume 45 to 65 per cent of total calories from carbohydrates. As carbohydrate is utilized as main source of energy, at least 40 per cent of the total energy in the food should come from Carbohydrates. In our country 60 – 80 percent of energy needs are met from carbohydrates in the form of starch furnished by cereals and pulses. It is recommended to consume less than 25 per cent of total energy from sugars. Sugar may displace nutrient rich foods from the diet, potentially leading to deficiencies of certain micronutrients. Consumption of excess amount of carbohydrate in the form of sugars may lead to dental carries and some other complications like diabetes, obesity *etc.*

Energy Value of Food

Energy is defined as the capacity for doing work. The energy contained within the chemical constituents of food can be either trapped in the body or used to produce heat or allow the body to move. So energy in nutrition is focused on chemical energy and the kinetic energy of motion.

The energy value of food is expressed in kilocalories. Kilo calorie can be defined as "heat required to raise the temperature of one kg of water through one degree Celsius".

Inter Conversion Factors

- 1 K.cal = 4.184 J
- 1000 K.cal = 4.184 KJ
- 1000 K.cal = 4.184 MJ
- 1KJ = 0.239 K.cal
- 1MJ = 239 K.cal
- 100 MJ = 23900 K.cal

Determination of Energy Value of Food

Bomb Calorimeter

The energy value of foods is usually determined using the instrument called bomb calorimeter.

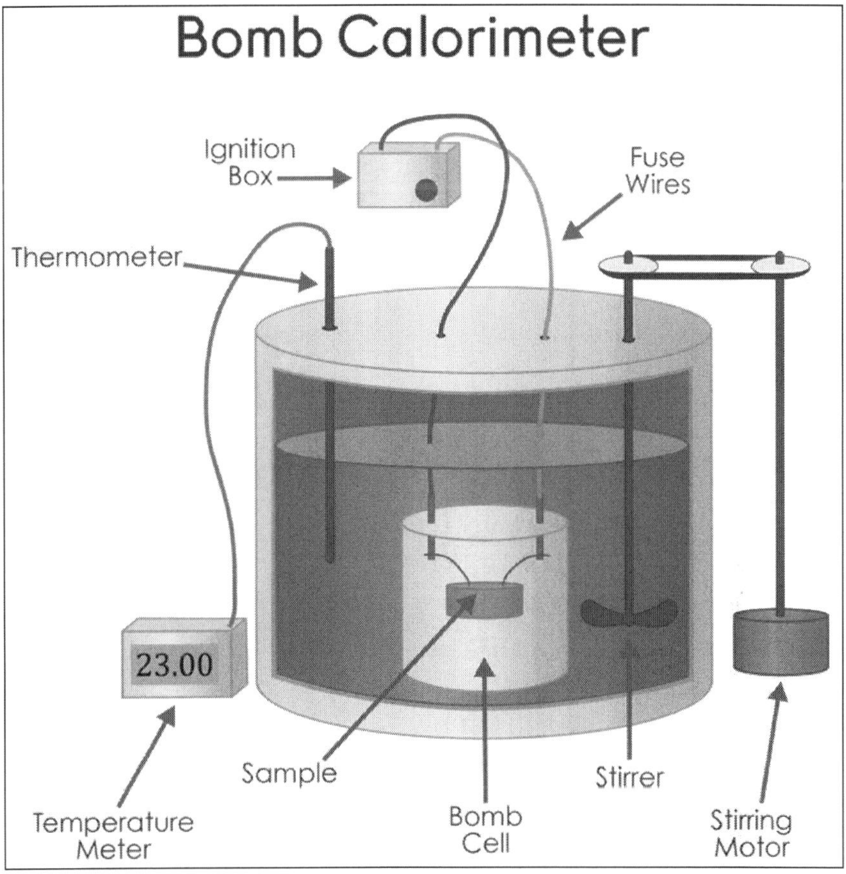

- It consists of a heavy steel bomb, with a platinum or gold- plated copper lining and a cover held tightly in place by means of a strong screw collar.
- A weighed amount of sample, usually pressed into pellet form, is placed in a capsule within the bomb which is then closed except for the oxygen valve, charged with oxygen to a pressure of about 300 pounds to the square inch.
- The oxygen valve is then closed and the bomb immersed in a weighed amount of water.
- The water is constantly stirred and its temperature taken at intervals of one minute by means of a differential thermometer, capable of being read to one thousandth of a degree.
- After the temperature of the water has been determined, the sample is ignited by means of an electric fuse and on account of the large amount of oxygen present; it undergoes rapid and complete combustion.
- The heat liberated is absorbed by the water in which the bomb is immersed and the resulting rise in temperature is accurately determined.

- The thermometer readings are also continued through an "after period", in order that the "radiation correction" may be calculated and the observed rise of temperature corrected accordingly.
- This corrected rise, multiplied by the total heat capacity of the apparatus and the water in which it is immersed, gives the total heat liberated in the bomb.
- From this, the heat arising from accessory combustions (the oxidation of the iron wire used as a fuse *etc.*) must be deducted to obtain the number of calories arising from the combustion of the sample.

Example

- Weight of sample taken = 2 g.
- Weight of water in the outside vessel = 3,000 g
- Water equivalent of the calorimeter = 500 g
- Initial temperature of water = 24° C.
- Final temperature of water = 26°C
- Rise in temperature = 2°C
- Heat gained by water and calorimeter = 3, 500 x 2 = 7,000 calories or 7Kcal
- 2 g sample produces =7 Kcal.
- 1 g sample produces = 3.5 Kcal.

Gross Fuel Value of Foods

The amount of energy release from the nutrients in bomb calorimeter or oxy calorimeter is known as gross fuel value.

- 1 g carbohydrate 4.1 Kcal
- 1 g fat 9.45 Kcal
- 1 g protein 5.65 Kcal

Physiological Fuel Value of Foods

The amount of energy actually available in the body from a given amount nutrient is known as physiological fuel value.

- 1 g carbohydrates 4.0 Kcal
- 1 g fat 9.0 Kcal
- 1 g protein 4.0 Kcal

Chapter 5

Protein: Classifications, Functions, Digestion, Absorption, Deficiency Symptoms, Sources, Requirements and Protein Quality

Dietary protein performs all three functions of nutrients. It is needed for growth, maintenance and repair of body tissue. The name protein was suggested by Mulder in 1838. One fifth of an adults total body weight is protein. He defined protein as nitrogen – containing constituents of foods and felt life was impossible without it. Protein is found in every cell of our body. All the tissues in our body such as muscle, blood, bone, skin and hair are made up of proteins. About 50 per cent of protein found in muscle, 20 per cent in bone, 10 per cent in skin and rest is present in other parts of the body.

All the proteins are synthesized from amino acid molecules and 22 different amino acids are used in protein synthesis. Amino acids contain carbon, hydrogen, oxygen and nitrogen atoms and cysteine and methionine contain sulphur atom. Unlike carbohydrates and lipids protein contains nitrogen atoms. All amino acid contain an Amino group (NH_2), a carboxyl group (COOH), and a hydrogen atom attached to a central carbon atom, which is also bonded to the side chains or side group of the amino acids.

$$\begin{array}{c} H \\ | \\ R - C - COOH \\ | \\ NH_2 \end{array}$$

Amino Acid structure

Classification of Protein

1. Based on number of amino acids and carboxylic acids
2. Based on the composition
3. Nutritional classification of protein

1. Based on Number of Amino Acids and Carboxylic Acids

Mono amino – mono carboxylic acids

These include glycine, alanine, valine, leucine, isoleucine, serine and threornine.

Mono amino – di carboxylic acids

These include aspartic acid and glutamic acid.

Diamino – mono carboxylic acids

Arginine and lysine belong to this, groups.

Sulphur containing amino acids

These are cystine, cysteine and methionine

Aromatic and heterocyclic amino acids

These include phenylalanine, tyrosine, histidine, tryptophan, praline and hydroxyl proline

2. Based on the Composition

Simple Protein

Proteins which yield only amino acids on hydrolysis are called as simple protein.

Example: albumins, globulins, glutelins, prolamins, histones, protamins and fibrous protein

$$\text{Simple Protein} \xrightarrow{\text{Hydrolysis}} \text{Amino acids}$$

Conjugated or Compound Proteins

The protein which combined with non-protein substances like carbohydrate, lipid, phosphorus, nucleic acid are called as conjugated protein.

Example: glycoprotein, lipoprotein, nucleoprotein, haemoglobins

- ☆ Glycoprotein- Protein and Carbohydrate
- ☆ Lipoprotein - Lipid and Protein
- ☆ Phosphoprotein - Phosphorus and Protein
- ☆ Haemoglobin - Haem and Protein

Derived Proteins

These proteins are derived by partial to complete hydrolysis of simple or conjugated proteins by the action of acids, alkalies or enzymes.

Types of Derived Proteins

- ☆ Primary derived proteins – Proteans, Metaprotein, Coagulated Protein (boiled egg)
- ☆ Secondary derived proteins – Proteoses, Peptones, Peptides

Protein ⟶ Proteoses ⟶ Peptones ⟶ Peptides ⟶ Amino Acids

3. Nutritional Classification of Protein

I. Complete Protein

It contains all essential amino acids. *e.g.* Egg protein.

II Partially Complete Protein

It is partially lacking in one or more essential amino acids, *e.g.*, Wheat protein (gluten).

III Incomplete Protein

It is completely lacking in one or more essential amino acids, *e.g.*, Zein, gelatin.

Nutritional Classification of Amino Acids

Amino acids are classified into 3 groups.

Essential Amino Acids

An essential amino acid is one that cannot be synthesized by the body to meet the physiological needs. So it should be supplied by the diet.

Non-essential Amino Acids

Non-essential amino acids are those can be synthesized by our body.

List of Amino Acids

Amino Acids	*Example*
Essential amino acids	Histidine (*only for infant*), Isoleucine, Leucine, Lysine, Methionine, Phenylalanine, Threonine, Tryptophan and Valine **(HILL,MP,TTV)**
Semi/conditionally essential amino acids	Arginine, Cysteine, Glycine, Proline and Tyrosine **(ACCG, PT)**
Non - essential amino acids	Alanine, Asparagine, Aspartic Acid, Glutamic Acid, Glutamine and Serine **(AAA, GAGS)**

Functions/Importance of Protein

1. Growth and Maintenance of Tissues

New growth, including building of muscles can occur only when an appropriate mixture of amino acids is available over and above the amount needed for the maintenance and repair of existing tissue. The vital process of cell division also depends on protein.

2. Enzymes are Proteins

Enzymes, including those involved in digestion *e.g.,* Pepsin, trypsin, amylase *etc.*, are made up of proteins. It acts as catalyze in digestion of foods and other biochemical reactions.

3. Hormones are Proteins

Hormones such as insulin, growth hormones *etc.*, are also made up of protein. It regulates metabolic functions of cells.

4. Proteins Required for Transmitting of Nerve Signal

It aids in transmitting nerve signals from one nerve cell to another. The amino acid tryptophan serve as the precursor for the vitamin niacin and for serotonin, a vital neurotransmitter that is involved in transmitting nerve signals from one nerve cell to another.

5. Proteins Required for Muscle Contraction

Contractile protein, myosin and actin are made up of regulate muscle contraction.

6. Transport of Nutrients

Proteins play important role in the transport of nutrients from the intestine across the intestinal wall to the blood, from the blood to the tissues of the body and across the membranes of the cells of the tissues. These transport and membrane bound carrier proteins are usually specific to one nutrient.

Retinol binding protein, for example, binds to and transports only retinol. Lipoproteins can transport many different lipid molecules.

7. **Protein Helps in the Regulation of Water Balance in our Body**

Fluid in our body is distributed in intracellular and extracellular compartments. The extra cellular is divided into the intercellular (between the cells) and intravascular (within the blood vessels) compartments. The balance between the compartments is achieved by dissolved proteins and dissolved ions (electrolytes) primarily sodium [potassium ions].

8. **Proteins Required for Maintenance of Appropriate pH**

Normal body process continuously produces acids and bases that must be carried by the blood to the organs of excretion. Proteins in the blood serve as a buffer. It will maintains the appropriate pH

9. **Antibodies are made up of proteins. It helps to fight against the infection.**

10. **Protein also a good source of energy. It provides 4 Kcal per gram of protein.**

Digestion of Protein

Stomach

Pepsin is present in gastric juice and hydrolyses the peptide bonds in protein molecules. It hydrolyses the dietary protein into polypeptides.

$$\text{Dietary protein} \xrightarrow{\text{Pepsin}} \text{Polypeptides}$$

Small Intestine

The polypeptides formed in stomach are digested in small intestine by trypsin, chymotrypsin secreted in the pancreatic juice. Polypeptides are peptides, dipeptides or tripeptides.

$$\text{Polypeptide} \xrightarrow{\text{Trypsin and Chymotrypsin}} \text{Peptides + Amino acids}$$

The carboxypeptidase hydrolyses the peptides in to amino acid.

$$\text{Peptides} \xrightarrow{\text{Carboxypeptidase}} \text{Amino acids}$$

The intestinal mucosa also contains dipeptidase and tripeptidase which hydrolyses dipeptides and tripeptides, respectively into amino acids.

$$\text{Dipeptides} \xrightarrow{\text{Dipeptidase}} \text{Amino acids}$$

$$\text{Tripeptides} \xrightarrow{\text{Tripeptidase}} \text{Amino acids}$$

Absorption

The end product of digestion of protein is amino acids. The amino acids are absorbed in small intestine. According to the available information, the amino acids associates with the carrier and Na ion in the microvilli and the complex travels to the inner side of the membrane where it dissociates, releasing the amino acid and Na ion into the cytoplasm. The carrier returns back and functions repeatedly. The Na ion is then actively transported out of the cell.

Deficiency Diseases

Protein Energy Malnutrition (PEM)/Protein Calorie Malnutrition (PCM)

Protein Energy Malnutrition (PEM) is a term used to describe clinical disorders resulting from varying degrees of protein and energy deficiency.

- Kwashiorkor
- Marasmus
- Marasmic Kwashiorkor

Kwashiorkor

The word Kwashiorkor was first introduced by Dr. Cicely Williams. The term 'Kwashiorkor' means 'the disease which the child gets when the next baby is born'.

Signs and Symptoms of Kwashiorkor

1. Growth failure and wasting of muscle
2. Behavioural changes like irritability and apathy are common
3. Oedema which is the accumulation of fluid in the tissues making them soft and spongy
4. Scaly pigmentation of skin (crazy pavement dermatosis)
5. Changes in hair (dry and sparse and red colour).
6. Loss of appetite, vomiting, diarrhoea
7. Enlargement of the liver
8. Anaemia
9. Moon face (full well rounded face)

Marasmus

Nutritional marasmus is due to the consumption of diet deficient in both protein and energy.

Signs and Symptoms

1. Growth retardation
2. Severe muscle wasting
3. Loss of subcutaneous fat
4. Child appear skin and bones
5. Irritability and apathy
6. Frequent diarrhoea
7. Dehydration
8. Oedema and fatty liver are absent

Marasmic Kwashiorkor

The Child displays the symptoms of both marasmus and kwashiorkor

Signs and Symptoms

1. Loss of subcutaneous fat - Marasmus
2. Oedema, hair and skin changes - Kwashiorkor
3. No moon face
4. Presence of moderate anaemia - vitamin deficiencies

Food Source of Protein

Sl.No.	*Food Stuff*	*Protein (per cent)*
1	Meat	18-20
2	Skimmed Milk Powder (Cow's)	33
3	Whole Milk Powder (Cow's)	26
4	Cheese	24
5	Egg	13
6	Soya Bean	43
7	Ground Nut	25
8	Peas, Dry	20
9	Bengal Gram Whole	17
10	Rice	7

Recommended Dietary Allowances (RDA) of Carbohydrate

The Indian Council of Medical Research recommends 1.0 g/kg/day as the safe level of intake in terms of dietary protein for Indians.

Group	*Particulars*	*Protein (g)*
Men	Sedentary work	60
	Moderate work	
	Heavy work	
Women	Sedentary work	55
	Moderate work	
	Heavy work	
	Pregnant women	+23
	Lactation 0-6 months	+19
	6-12 months	+13
Infants	0-6 months	1.16 g/kg/d
	6-12 months	1.16 g/kg/d
Children	1-3 years (Preschool children)	16.7
	4-6 years	20.1
	7-9 years	29.5
Boys	10-12 years	39.9
Girls	10-12 years	40.4
Boys	13-15 years	54.3
Girls	13-15 years	51.9
Adolescents	16-17(Boys)	61.5
	16-17(Girls)	55.5

Protein Quality

The quality of a protein is determined by the kind and proportion of amino acid it contains. Proteins that contain all essential amino acids in proportions capable of promoting growth are described as complete protein, good quality protein, or proteins of high biological value.

A good quality protein is digested and utilized well. Egg protein is a complete protein and is considered as a reference protein with the highest biological value. The quality of other proteins is determined based on their comparison with egg protein.

The protein of animal foods like milk, meat, and fish generally compare well with egg in the essential amino acid composition and are categorized as good quality proteins.

Plant proteins are of poor quality, since the essential amino acid composition is not well balanced.

PER, NPR, NPU and amino acid score are most suitable methods of evaluation of quality of protein.

Digestibility Coefficient

It refers to the percentage of the ingested protein absorbed into the blood stream after the digestion is completed.

Albino rats (28 days old)

I Group — II Group

Fed protein diet — Fed protein free diet

Determine faecal nitrogen — Determine endogenous faecal nitrogen

Calculate the Digestibility coefficient

$$\text{Digestibility coefficient} = \frac{\text{Nitrogen intake (g)} - \text{Nitrogen loss in digestion (g)}}{\text{Nitrogen intake (g)}} \times 100$$

Biological Value

Biological value of protein is the percentage of nitrogen that is absorbed and available for use by the body for growth and maintenance. It measures the amount of dietary protein utilised by body for maintenance and growth. The method is briefly as follows:

Albino rats (28 days old)

I Group — II Group

Fed test diet (10 per cent protein) Fed non-protein diet

↓

Record the food intake

↓

Collect urine and faeces

↓

Determine the nitrogen content

↓

Calculate the NPU

$$\text{Biological value} = \frac{\text{Nitrogen digested} - \text{Nitrogen lost in metabolism}}{\text{Nitrogen digested}} \times 100$$

Net Protein Utilization

Miller and bender (1955) developed a direct method of estimating 'Net Protein Utilisation'. The net protein utilization takes into account the losses in the digestion and metabolism.

$$\text{N.P.U} = \frac{\text{Digestibility co-efficient X biological value}}{100}$$

Protein Efficiency Ratio (PER)

Protein efficiency ratio is defined as the weight gain per gram of protein intake. This method was developed by Osborne, Mendel and Ferry.

$$\text{PER} = \frac{\text{Gain in weight (g)}}{\text{Protein intake (g)}}$$

Amino Acid Score/Chemical Score

The amino acid score is the ratio between the content of the most limiting amino acid in the test protein to the content of the same amino acid in egg protein (reference protein).

$$\text{Amino acid score} = \frac{\text{Limiting amino acid content in the test protein}}{\text{The same amino acid in egg protein}} \times 100$$

Chapter

6

Fat: Classification, Functions, Digestion and Absorption, Deficiency Symptoms, Sources, Requirements and Rancidity

The word lipid is derived from the Greek word '*lipos*' meaning fat. They are Group of naturally occurring substances characterized by their insolubility in water, greasy feel and solubility in some organic solvents. *e.g.* ether, chloroform, boiling alcohol and benzene. Like carbohydrates, fats also contain carbon, hydrogen and oxygen. Fats are the main energy store of the body. Fats are known as triglycerides formed by the combination of 3 molecules of fatty acids linked with glycerol. On oxidation, fats give almost two and a half times the energy as that of glucose or glycogen. Due to the lower amount of oxygen fat being a more concentrated source of energy (1g of fat = 9kcalories).

Fat are generally solids at room temperature. *e.g.* Vanaspathi, butter

Oils are liquid at room temperature. *e.g.* coconut oil, groundnut oil

Lipids are applied to group of naturally occurring substances characterized by their insolubility in water, greasy feel and solubility in some organic solvents.

Classification of Lipids

Lipids

Simple lipids	**Compound lipids**	**Derived lipids**
Fats and oils	Phospholipids	Fatty acids
Waxes	Glycolipids	Alcohol
	Aminolipids	Steroids
	Sulpholipids	

1. Simple Lipids

a) Fats and Oils

- ✰ Fats and oils are esters of fatty acids with glycerol.
- ✰ They are also called as triglycerides
- ✰ Oils are liquids at 20°C, fats are solid at 20°C.
- ✰ Lipids are insoluble in water, soluble in organic solvents like ether, chloroform, acetone and benzene. *e.g.*, vegetable oil, ghee, butter, lard

$$\text{Fats} \longrightarrow \text{Fatty Acid + Glycerol}$$

b) Waxes

- ✰ Waxes are esters of fatty acids and alcohol.eg long chain aliphatic alcohols or with cyclic alcohols.
- ✰ In human body it is present in the form of esters of cholesterol

  E.g. Plant wax, insect wax (bees wax)

$$\text{Waxes} \longrightarrow \text{Fatty Acid + Alcohol}$$

2. Compound Lipids

The compound lipids contain, in addition to fatty acids and glycerol, some other organic compounds.

a) Phospholipids

These contain phosphoric acid and nitrogenous base in addition to fatty acids and glycerol. *e.g.* lecithin and cephalin

$$\text{Phospholipids} \longrightarrow \text{Fatty Acid + Glycerol + Phosphoric acid}$$

b) Glycolipids

These lipids contain carbohydrates in combination with fatty acids and glycerol.

e.g. cerebrosides

Glycolipids ⟶ Fatty Acid + Glycerol + Galactose

c) Lipoprotein

It contains protein subunits along with lipids.

Lipoprotein ⟶ Lipid + Protein

3. Derived Lipids

Substances derived from simple and compound lipids by hydrolysis are called as derived lipids. Fatty acids, alcohols and sterols are example for derived lipids.

a) Fatty Acids

Fatty acids are obtained by hydrolysis of fat. There are 40 different fatty acids are found in nature. The common formula is $CH_3(CH_2)nCOOH$, where n can be any even number from 2-24.

Based on number of double bonds between carbon atoms fatty acids are classified as **Saturated fatty acids** and **unsaturated fatty acids**

- Saturated fatty acids are classified into short chain (4-8) fatty acids, medium chain (10-14) fatty acids, long chain (16-22) fatty acids
- Unsaturated fatty acids divided into monounsaturated fatty acids,polyunsaturated fatty acids

Saturated Fatty Acids

Fatty Acids	*C Atoms and Double Bonds*	*Occurrence*
Short chain fatty acids		
B-Butric acid	C4 : 0	Butter
C- caproic acid	C6 : 0	Butter, Coconut oil
C- Caprylic acid	C8 : 0	Butter, Coconut oil
Medium chain fatty acids		
C- Capric acid	C10 : 0	Butter, Coconut oil
L- Lauric acid	C12 : 0	Butter, Coconut oil
M-Myristic acid	C14 : 0	Butter, Vegetable foods
Long chain fatty acids		
P- Palmitic acid	C16 : 0	Most vegetable and animal fats
S- stearic acid	C18 : 0	Most vegetable and animal fats
A- Arachidic acid	C20 : 0	Butter, lard, peanut oil
B- Behenic acid	C22 : 0	Vegetable oils

Unsaturated Fatty Acids

Fatty Acids	*C Atoms and Double Bonds*	*Occurrence*
Monounsaturated fatty acids		
P- Palmitoleic acid	C16 :1	Olive oil, fish oil, beef oil
O- Oleic acid	C18 : 1	Olive oil, canola oil
E- Erucic acid	C22 : 1	Rape seed oil, canola oil
Polyunsaturated fatty acids		
L- Linoleic acid	C18: 2(n-6) ω-6 fatty acid	Vegetable seed oils (safflower, corn, soya been, cotton seed)
α linolenic acid	C18 : 3(n-3) ω-3 fatty acid	Vegetable seed oil (soya been oil)
Archidonic acid	C20 : 4 (n-6) ω-6 fatty acid	Liver, lard, meat
Eicosapentaenoic acid	C20 : 5(n-3) ω-3 fatty acid	Fish oils, shell fish
Docosahexaenoic acid	C22 : 6(n-3) ω-3 fatty acid	Fish oils, shell fish

Essential Fatty Acids

Essential fatty acids cannot be synthesized in the body. They are derived only from food. The essential fatty acids necessary for growth are linoleic acid, linolenic acid and arachidonic acid.

Non-Essential Fatty Acids

Non-essential fats are those that are synthesized by the body and are not necessary in the diet.

b) Alcohols

Alcohol includes glycerol, cholesterol and higher alcohol.

c) Sterols

Cholesterol and ergosterol are some the sterols associated with fat.

Functions of Fat

- ☆ It is a concentrated source of energy. One gram of carbohydrate gives 9Kcal.
- ☆ Fats are essential for the absorption of fat soluble vitamins like A, D, E, K.

- Fats contain essential fatty acids like linolic, linolenic and arachidonic acids which are essential for maintaining tissues in normal health.
    - Vegetable oils are good source of vitamin E.
    - Butter and fish liver oil are good source of vitamin A.
- Fats improve the palatability of the diet
- It gives satiety value, *i.e.*, a feeling of fullness in the stomach
- Phospholipids and other compound lipids are essential constituents of nervous tissue.
- Fats are deposited in the adipose tissue and this deposit serves as the source of energy during starvation.
- Fat deposited in adipose tissue functions like an insulating material against cold and physical injury.
- Cholesterol is essential for maintaining the cell membrane in good condition.
- Cholesterol is act as a precursor for the formation of some steroid hormones
- Small amount of fat in the diet is essential to provide the omega-3 and omega -6 fatty acids to the body. At least 15 per cent of the total energy should be supplied through fat source.

Deficiency

- Severely low fat diet can result in poor vitamin (fat soluble) absorption
- Depression
- An increased risk of cancer
- Dry skin (phrenoderma)
- Increased susceptibility to infections and problem with bone and tooth development.
- Cognitive problem
- Vision problem
- Arthritis
- Cracked nails
- Dry and lifeless hair
- Dry mucous membranes (in the tear ducts, mouth, vagina, and other areas)
- Fatigue, malaise, and low energy.

Digestion and Absorption of Fats

Stomach

Lipase present in stomach is unable to digest fats due to high acidity of gastric juice. So fat is not digested in stomach.

Small Intestine

Bile salt is essential for the digestion and absorption of fat. It helps to emulsify fats before digestion.

Pancreatic lipase and intestinal lipase hydrolyze the fat into diglycerides, monoglycerides and fatty acids.

Phospholipase and cholesterol esterase hydrolyze phospholipids and cholesterol

Fat Digestion

Fat (triglycerides)

↓ Bile salt

Emulsification

Pancreatic lipase ↓ Intestinal lipase

Monoglycerides and diglycerides

Absorption of Fats

The products of digestion pass into the cells of the intestinal wall, where synthesis of new glycerides takes place. The lipids pass through the lacteals of the small intestine to thoracic duct and then to the blood stream in the form of fine particles known as chylomicrons. A greater part of the cholesterol present in the diet is absorbed while phytosterols present in vegetable fats and oils are not absorbed.

Lipids in Blood

Normal human plasma in the post absorption state contains about 500 mg of total lipids (per100ml) of which about 120 mg are triglycerides, 160 mg phospholipids, 180 mg cholesterol and about 10-15 mg free fatty acids. Plasma contains two lipoproteins which are involved in the transport of fat.

Storage of Fat

Fat is stored in the adipose tissues in normal human subjects, adipose tissue constitutes about 10-15 per cent of the body weight. It increases up to 30 percent in obese person.

Food Sources

Animal sources: Animal sources of fats are meat, fish, eggs, milk, ghee and butter.

Vegetable sources: They are vegetable oils obtained from ground nut, sesame, coconut and mustard.

Other sources: Minute quantities of fat are found in other foods such as cereals, pulses, nuts and vegetables. This fat is called invisible fat.

RDA of Fat

Sl.No.	*Group*	*Requirement (g/day)*	*Total Calories from Fat (per cent)*
1	Infants (0-1 year)	19	25-30
2	Children	25-30	15-20
3	**Adolescents (13-17 year)** : Boys	45-50	
	Girls	35-40	
4	Man	25-40	10-20
5	Woman	20-30	

Deficiency of Fats

During infancy deficiency of essential fatty acids results in perianal irritation and skin changes like dry skin and itching. In adults phrynoderma is easily cured by essential fatty acids with vitamins B_2 and A.

Rancidity

Rancidity typically refers to the release of aldehydes and ketones in fats that have unpleasant taste and odors. The development of off-flavours or unpleasant odour in fats is known as rancidity. There are three main types of rancidity. Fats containing unsaturated fatty acids are more susceptible to rancidity.

1. Hydrolytic rancidity
2. Oxidative rancidity and
3. Microbial rancidity

Hydrolytic Rancidity

- In the presence of moisture in fatty foods triglycerides go through the process of hydrolysis, breaking them into their fatty acids and glycerol.
- These individual fatty acids are more easily distinguishable by taste and typically have an unpleasant taste and odor

Oxidative Rancidity

When these types of oils or fats are exposed to oxidation, they begin decomposing into short-chain fatty acids, such as butyric acid, which gives the fats a rancid taste

Microbial Rancidity

It is caused by bacteria breaking down fats with enzymes, causing the fat to decompose.

Chapter 7

Fat Soluble Vitamins: Functions, Deficiency Symptoms, Sources and Requirements

Vitamins are "vital (essential) for life" The term vitamin was coined by **Funk** in 1912. "Vitamins may be defined as tasteless, organic compounds occurring in small quantities in the different natural foods and necessary for the growth and maintenance of good health in human beings." They are involved in the utilization of the major nutrients like proteins, fats and carbohydrates.

Classification of Vitamins

Vitamins

Fat soluble	**Water soluble**
(Vitamin A, D, E, K)	(Vitamin B and Vitamin C)

Fat soluble vitamins are soluble in fat and stored in the body. Vitamin A, D, E, and K are fat soluble vitamins.

Vitamin A

Vitamin A is discovered in the year of 1909. The chemical name of vitamin a is **retinol.** Vitamin A is a generic term for many related compounds. Vitamin A activity is also possessed by carotenoids present in plants. So carotenoids (β- carotene) are called as provitamin A or precursor of vitamin A.

Provitamins - substances in foods that are not directly usable by the body.

- ☆ Converted to an active form once absorbed
- ☆ Example: **beta-carotene**

Preformed Vitamins

- ☆ Vitamins in foods in their active form.
- ☆ Do not need to be converted.
- ☆ Retinol (alcohol), Retinal (aldehyde) are often called preformed vitamin A.

Functions

1. Vitamin A is very important for good vision.

   In vision retinol supplied to the retina in blood is converted to retinal, which then combines with a protein called **opsin** to produce the purple pigment known as rhodopsin. Rhodopsin also known as visual purple located in the light sensitive rod cells of the retina. When light strikes the retina, the rod cells are bleached as rhodopsin splits to form retinal and opsin. As this occurs a nerve stimulus is transmitted through the optic nerve fibres to the visual central of the brain, where the sensation of vision is created.

The Visual Cycle

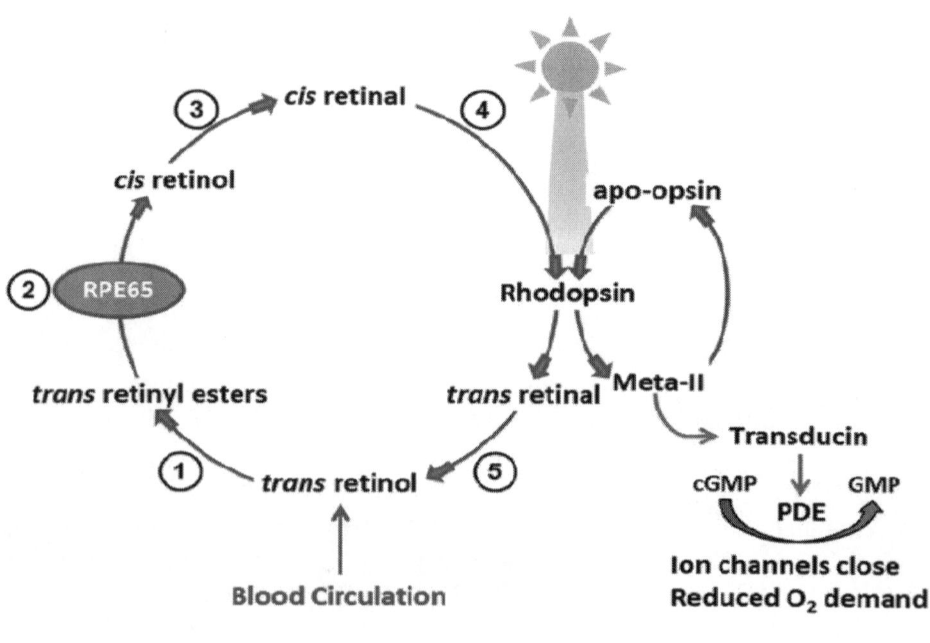

2. Maintaining the health of the epithelial tissues of eyes and skin.
3. Supports reproduction and growth.

   In vitamin A deficiency, reproduction does not take place through

   (i) in fertility in male and

   (ii) failure of the female to conceive or abortion of the foetus if conceived.

   An early symptom of vitamin A deficiency is loss of appetite, followed by a cessation of growth, then rapid weight loss and ultimate death.
4. It is essential for normal bone formation.

   It is essential for normal growth of bones. In vitamin A deficiency bone becomes weak, although thicker than normal bones. The cavities in the skull and spinal column do not enlarge to make room for the growing nervous system.
5. Vitamin A essential for mucoprotein and glycoprotein synthesis
6. It helps to strengthen immune system.
7. Beta carotene acts as antioxidant and protecting the body against cancer.

Deficiency Diseases

Night Blindness

Inability to see object in dim light due to vitamin A deficiency is called night blindness. When vitamin A is deficient the formation of rhodopsin is impaired giving rise to night blindness.

Bitots Spots

It is more an extension of xerotic process. These spots are raised, muddy and dry triangular batches. It occurs in older children's or young adults

Xerosis of Conjunctiva

It manifests as dry patches of non wettable conjunctiva. It may be associate with various degrees of thickening, wrinkling and pigmentation (muddy coloring) of the conjunctiva. Children's under the 5 years of age are affected due to deficiency.

Xerosis of cornea-when dryness spread to the cornea there is dull, hazy, lustreless appearance.

Keratomalacia

Ulceration of conjunctiva or cornea is called Keratomalacia. In this condition softening and dissolution of the cornea occurs. If the process is not treated, perforation of cornea lead to prolapse the iris, extrusion of lens and infection to whole eyeball occur.

World Health Organization (WHO) Classification of Vitamin A Deficiency

Grade of xerophthalmia		Peak age group (years)	Type of deficiency
XN	Night blindness	2–6; adult women	Longstanding. Not blinding
X1A	Conjunctival xerosis	3–6	Longstanding. Not blinding
X1B	Bitot's spot	3–6	Longstanding. Not blinding
X2	Corneal xerosis	1–4	Acute deficiency. Can be blinding
X3A	Corneal ulcer/ <1/3 cornea	1–4	Severe acute deficiency. Blinding
X3B	Corneal ulcer/keratomalacia ≥1/3	1–4	Severe acute deficiency. Blinding
XS	Corneal scarring (from X3)	>2	Consequence of corneal ulceration
XF	Xerophthalmic fundus	Adults	Longstanding. Not blinding. Rare

Sources

Sheep liver, butter, ghee, egg, milk, curds, liver oils of shark, green leafy vegetables, carrots, mango and papaya are the rich source of vitamin A.

Requirements/RDA of Vitamin A

Age Group	*Requirements (µg)*
Children	400
Adolescents and adult	600
Pregnancy	800
Lactation	950

International units (IU) of vitamin A- 0.3 µg

International units (IU) of β carotene - 0.6 µg

1 µg of vitamin A-4 µg β carotene

Vitamin D (Calciferol)

Vitamin D can be synthesized in the body in adequate amounts by simple exposure to sunlight, even for 5 minutes per day is sufficient. The **7-dehydrocholesterol** is present in skin and act as **precursor of vitamin D**. It is essential for bone growth and calcium metabolism. It acts as a hormone in the body by facilitating calcium absorption and deposition in the bone.

Forms of Vitamin D

Vitamin D2 - Ergocalciferol

Vitamin D3 – Cholecalciferol

Functions

1. Vitamin D helps in the absorption of calcium and phosphorous by facilitating the increase the level of calcium binding protein in intestine. When vitamin D is absent, absorption of calcium is very low and greater part of the ingested calcium is lost in the faeces.
2. It helps in the formation of bone and teeth.
3. Vitamin D helps to maintain the calcium and phosphorous levels in the body

Sources

Sunlight, Cod liver oil, shrimp, liver, butter, yolk, cheese, milk, spinach and cabbage

Vitamin D Metabolism

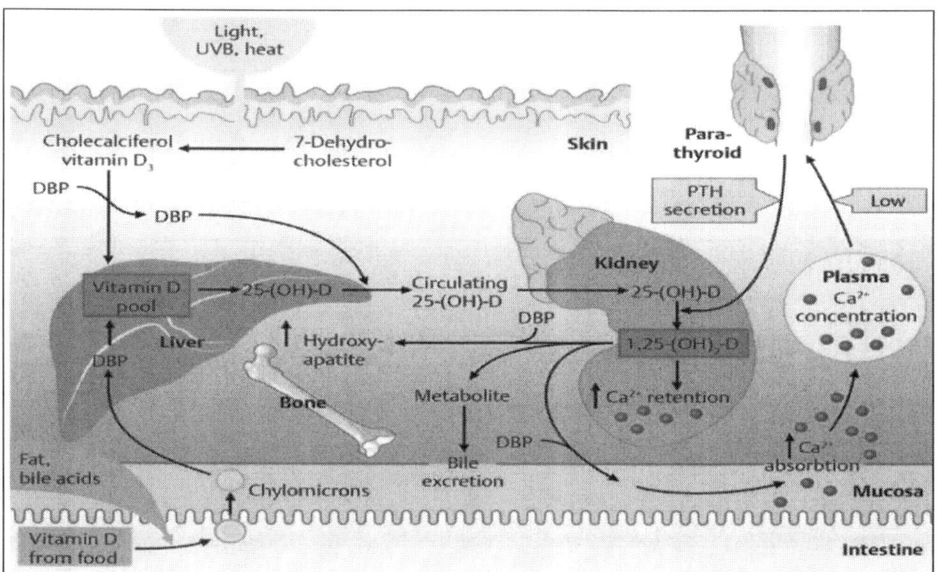

DBP (Vitamin D binding protein)

Deficiency Diseases

Rickets (Children)

The weakness and abnormalities in bone formation due to vitamin D deficiency is called rickets. Rickets primarily affects children.

Signs and Symptoms

- ✰ Enlarged and elongated and flattened head (box head)
- ✰ Pigeon breast
- ✰ Swelling in wrist and ankles
- ✰ Deformities in pelvis
- ✰ Flat feet
- ✰ Pot belly

Osteomalacia (Adult rickets)

The unmineralised condition of bone (soft bone) is called Osteomalacia. It occurs in women in *Purdah*, not exposed to sunshine.

Signs and Symptoms

- ✰ Deformities in pelvis and leg
- ✰ Pain in pelvis, lower back, legs and
- ✰ Spontaneous bone fractures.

Osteoporosis (old age)

- ✰ Bone pain
- ✰ Bone fracture

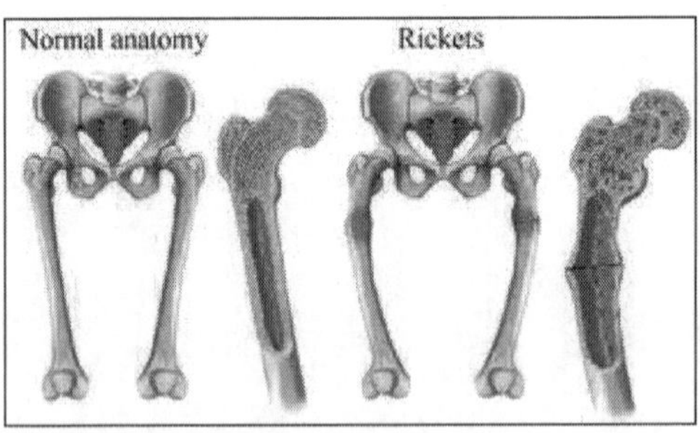

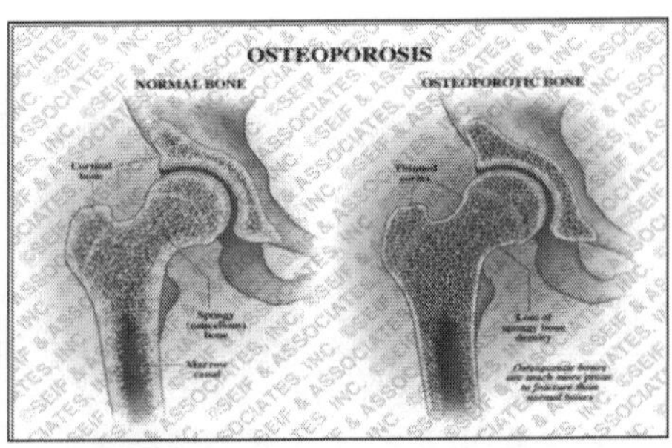

Requirements/RDA of Vitamin D

Age Group	*Requirements (mg)*
Infant and preschool children	10
Older children and Adult	5
Expectant and Nursing mother	10

Vitamin E (Tocopherol/Antisterility vitamin)

Vitamin E is known as 'antisterility vitamin' because it is required for normal reproduction in animals and men. The term tocopherol derived from *Greek* words '*Tocos*' (children), '*Phero*' (to bear) and '*Ol*' (alcohol).

Functions

1. Vitamin E is helps in maintains integrity of the cell membrane by preventing oxidation of poly unsaturated fatty acids (PUFA).
2. It also prevents the oxidation of β- carotene and Vitamin A.
3. It protects RBC against haemolysis by oxidising agents.
4. Vitamin E reduces platelet aggregation.
5. Vitamin E is essential for the maintenance of immune function.
6. Vitamin E prevents aging process.

Food Sources

Wheat germ, Vegetable oils, sunflower seeds, almonds, safflower oil, Sesame oil and mustard oil, eggs, butter are the rich source of vitamin E.

Deficiency Diseases

Reproductive Failures

In female rat fed on vitamin E deficient diets, foetal death and abortion take place. The primary metabolic defect which causes death and subsequent abortion of the embryo is now known. There are degenerative changes in the uterus. However, the vascular system of the embryo also undergoes degeneration. In the male rat, irreversible testicular degeneration occurs in vitamin deficiency.

- ☆ Liver necrosis
- ☆ Muscular dystrophy
- ☆ Haemolytic anaemia
- ☆ Haemorrhagic disease of the new born

Requirements/RDA of Vitamin E

Age Group	*Requirements (mg)*
Infant and preschool children	3-4
Adult	10
Pregnancy	12
Lactation	13

Vitamin K (Antihaemorrhagic)

Vitamin K is recognized as the antihaemorrhagic factor due to its vital role in blood clotting mechanism.

Forms of Vitamin K

Vitamin K1- Phyloquinone

Vitamin K2- Menaquinone

Vitamin K3 - Menadione

Functions

It aids in blood clotting. Vitamin K is essential for the activation of prothrombin. This gets converted to thrombin, which in turn activates fibrinogen to form fibrin.

Food Sources

Dark green leafy vegetables, cabbage, Fruits, tubers, seeds, dairy and meat products are the rich source of vitamin K.

Effects of Deficiency

Vitamin K deficiency leads to lowering of prothrombin level and increased clotting time of blood. This may lead to haemorrhagic conditions. Vitamin K deficiency occurs in the following ways:

1. Inadequate intake of vitamin K
2. Inadequate intestinal absorption

Inadequate Intake of Vitamin K

Inadequate intake of vitamin K by the mother may causes the haemorrhagic disease of the new born. The infants have low prothrombin level and they recover rapidly whenever Vitamin K is administered by injection.

Inadequate Intestinal Absorption

It results from lack of bile in the intestine due to defective secretion of bile as in liver disorders intestinal obstruction and poor absorption due to diarrhoea or dysentery.

Requirements/RDA of Vitamin K

Age Group	*Requirements (mg)*
Infant	12-20
Adult	70-140
Pregnancy	200
Lactation	150
Adolescents	50-100

Chapter

8

Water Soluble Vitamins: Functions, Deficiency Symptoms, Sources and Requirements

Water-soluble vitamins are dissolved in water which is not stored in the body; they are eliminated through urine. We need a continuous supply of them in our diets. The water-soluble vitamins are the B-complex group and vitamin C. Fat-soluble vitamins - vitamins A, D, E and K — dissolve in fat before they are absorbed in the blood stream to carry out their functions. Excesses of these vitamins are stored in the liver. Because they are stored, they are not needed every day in the diet.

Water-soluble vitamins are easily destroyed or washed out during food storage or preparation. Proper storage and preparation of food can minimize vitamin loss. To reduce vitamin loss:

1. Refrigerate fresh produce; eat raw if possible
2. Keep milk and grains away from strong light
3. Use as little water as possible when cooking vegetables
4. Use a lid when cooking to prevent evaporation of cooking liquids
5. And use the cooking water from vegetables to prepare soups
6. Avoid alkalinity during cooking such as the use of baking soda
7. Use whole grains, with as little processing as possible

Vitamin C (Ascorbic Acid)

The chemical name for Vitamin C is ascorbic acid. It was discovered in 1747 by the British physician Lind and demonstrated that citrus fruit juices prevented and cured scurvy. It contains six carbon atom related to the monosaccharide glucose.

Vitamin C is stable to acid, destroyed by oxidation, light, alkali, heat especially in the presence of iron and copper. Oxidised form of ascorbic acid is dehydroascorbic acid and it have vitamin C activity. Most of the animal synthesis vitamin C from glucose but a few including humans lack the liver enzyme gulanolactone-oxidase, which is required to catalyse one step of this process. In plants vitamin C accumulates during ripening process and is presumably synthesised within the plant cells from naturally occurring glucose.

Functions

1. Ascorbic Acid is essential for formation of collagen which is the structural protein of connective tissues, blood vessels teeth and bones.
2. It helps in the biosynthesis of non-essential amino acids.
3. Able to keep ferric ions in ferrous form and facilitate the absorption
4. It is required for absorption of iron.
5. It is required for normal wound healing.
6. Vitamin C is an excellent anti-oxidant and prevents oxidation of cells.
7. Vitamin C is required for carnitine synthesis which helps in the transport of fatty acids in the cell.
8. Vitamin C is essential for the synthesis of norepinephrine a neurotransmitter.
9. It activates hormones.
10. Drug detoxifying metabolic systems in the body requires vitamin C for its optimal activity of various drug detoxifying metabolic system with the body.

Food Sources

Amla, drumstick leaves, guava, cashew fruit, agathi, cabbage, bitter gourd, oranges, and tomatoes are the rich source of vitamin C.

Deficiency Diseases and Symptoms

Infantile Scurvy

1. Pain and tenderness of the limbs
2. Swelling on limb
3. Purpuric spots under the skin
4. Gums are swollen and purplish

Adult Scurvy

1. Bleeding gums
2. Weakness and fatigue
3. Haemorrhages in calf
4. Pain in bones
5. Swelling and pain in joints and muscles

Delayed Wound Healing

Requirements/RDA of Vitamin C

Age Group	*Requirements (mg)*
Infant	25
Preschool children	40
Adolescents and Adult	40
Pregnancy	60
Lactation	80

Thiamine (Vitamin B_1)

Thiamine is known as Vitamin B_1. Deficiency of thiamine leads to beri – beri. This condition is widely prevalent among population whose diet contains more of polished cereals.

Functions

1. Thiamine is converted to thiamine pyrophosphate (TPP) when 2 phosphate groups are added to the basic structure of thiamine, which is an important for the carbohydrate metabolism.TPP is required for decarboxylation of pyruvate (pyruvic acid) to form acetyl CO A and for α-ketaglutarate to succinyl CO A. The pyruvate and α-ketaglutarate accumulate in those with thiamine deficiency.
2. It is involved in transmission of nerve impulses across the cells
3. Thiamine as TPP is an essential cofactor for the conversion of amino acid tryptophan to niacin.

Sources

Yeast, whole wheat, millets, hand pounded rice, parboiled rice, bran, gingelly seeds, groundnut, soya bean, cashew nuts, organ meats, pork, liver and eggs.

Deficiency Diseases

- ☆ Loss of appetite and anorexia are the first symptoms of thiamine deficiency.
- ☆ Degreased muscle tone in the gastro intestinal tract results in degreased gastric motility, distressed colon, and constipation
- ☆ **Dry beriberi** – Wasting of muscles, difficult to walk
- ☆ **Wet beriberi** –due to the lack of thiamine glucose not metabolised and pyruvic acid accumulate in the tissues and body fluids. These metabolites cause dilation of peripheral blood vessels in beriberi. This vasodilation may be extreme so that fluid leaks out through capillaries produce **odema** Oedema, swelling in calf muscles, high diastolic pressure. At the same time blood flow rapidly through the dilated peripheral circulation. There is a high cardiac output.
- ☆ Oedema not only on leg but also in face trunk, serous cavities.
- ☆ **Infantile beriberi** – Infants of 3-5 years of age are affected by this disease. This condition develops quickly if not treated lead to death.
- ☆ **Cyanosis**- due to lack of oxygen or accumulation of CO_2 in the blood causes skin to turn blue.
- ☆ Enlargement of heart, Enlarged liver, tenderness of calf muscles

Requirements/RDA of Vitamin B_1

Age Group	*Requirements (mg)*
Infant	0.2-0.3
Preschool children	0.5
Adolescents and Adult	1.0-1.7
Pregnancy	1.6
Lactation	1.7

Riboflavin (Vitamin B_2)

Riboflavin or Vitamin B_2 is the yellow enzyme which is heat stable unlike other B Vitamins. It is unstable to alkali and light. Slightly soluble in water-some losses occurs when cooking with large amount of water for long period. Riboflavin in the combined form with proteins forms flavoproteins or yellow enzymes. Major losses of riboflavin is due to sun light upto 70 per cent when 4 hours exposed to sunlight.

Functions

1. Riboflavin is required for produce two coenzymes - FMN (Flavin Momo Nucleotide), FAD (Flavin Adenine Dinucleotide).

2. These coenzymes functions in many biological reactions primarily in oxidation-reduction and dehydrogenation reaction in electron transport transport chain link cycle because of their ability to accept and transfer hydrogen atoms.
3. It is required for the metabolism of carbohydrate, protein and fat.
4. It is required for the Conversion of vitamin B6 and folate to active coenzymes.
5. It is essential for the formation of red blood cells.
6. It is required for the synthesis of glycogen.

Food Sources

Rich sources are liver, dried yeast, egg, milk, meat, fish, whole cereals, legumes and green leafy vegetables.

Deficiency Diseases

Angular stomatitis, glositis, cheilosis are the characteristic features of riboflavin deficiency. Early symptoms are sourness and burning of mouth.

Angular Stomatitis

Lesions at the angles of the mouth is called Angular Stomatitis

Glossitis

The inflammation of the tongue with magenta colour is called glossitis

Cheilosis

Dry chapped appearance of the lip with ulcers is called cheilosis.

Scrotal Lesion

Dermatitis involving the facial and scrotal skin is common in riboflavin deficiency.

Ocular Lesions

In humans, visual symptoms photophobia, lacrimation, burning, visual fatigue is reported in riboflavin deficiency.

Requirements/RDA of Vitamin B_2

Age Group	*Requirements (mg)*
Infant	0.3 - 0.4
Preschool children	0.6
Adolescents	1.2
Adult	1.4 - 2.1
Pregnancy	2.1
Lactation	2.0

Niacin (Vitamin B_3)

Niacin or Nicotinamide (amide form) is required by all the cells of our body. It is originally obtained by oxidation of nicotine. Niacin is the common term for the derivatives of nicotinamide. Like thiamine and riboflavin it plays a vital role in the release of energy from carbohydrates protein, fat and alcohol. It is white crystalline substance readily soluble in water.

Functions

1. Like thiamine and riboflavin plays important role in release of energy from carbohydrates protein, fat, and alcohol.
2. Niacin is essential for tissue metabolism.
3. Required for synthesis of protein, fat and five carbon sugar(pentose) need for the synthesise of DNA and RNA
4. Essential for the normal skin, intestinal tract and nervous system
5. NAD and NADP are involved as coenzymes in large number of oxidation reduction reactions.
6. Nicotinic acid enhances stomach secretion.
7. NAD is involved in catabolic and NADP involved in anabolic reaction in our body.

Food Sources

Dried yeast, liver, rice polishing, peanut, whole cereals, legumes, meat and fish.

Effects of Deficiency

Deficiency of nicotinic acid causes a disease known as **pellagra**. It is characterized by **Three D diseases**.

1. **D**ermatitis (Rough skin)
2. **D**iarrhoea
3. **D**ementia (Irritability, depression, poor concentration and loss of memory)

Requirements/RDA of Vitamin B_3

Age Group	*Requirements (mg)*
Infant	650-710 µg/Kg
Preschool children	8
Adolescents	14
Adult	16-21
Pregnancy	16
Lactation	20

Folic Acid

Folic acid is the general term for the chemical – Pteroylglutamic acid and related compounds with the same biological activity as PGA. Folic acid was first extracted from dark green leafy vegetables. It forms yellow crystals and is a conjugated substance made up of three acids namely pteroic, para amino benzoic acid and glutamic acid.

Functions

1. It helps in the formation of haem group of haemoglobin.
2. It is essential for normal growth and division of cells.
3. It is essential for the maturation of red blood cells.
4. It is act as key compounds for the formation of Purines which are essential constituents of living cells.
5. Along with vitamin B12 – helps in transmethylation of homocystein to methionine, ethanolamine to choline and uracil to thymin.
4. It is essential for the formation Thymine which is essential compound of DNA.
5. It is necessary for the conversion of phenylalanine into tyrosine.

Food Sources

Green leafy vegetables, liver, kidney, gingelly seeds, cluster beans

Deficiency Diseases

1. **Megaloblastic anaemia**. In which condition red blood cells are large in size and low in volume. Megaloplast appear in bone marrow as well as in peripheral blood.
2. Neural tube disorders of the foetus.
3. Impaired immune system

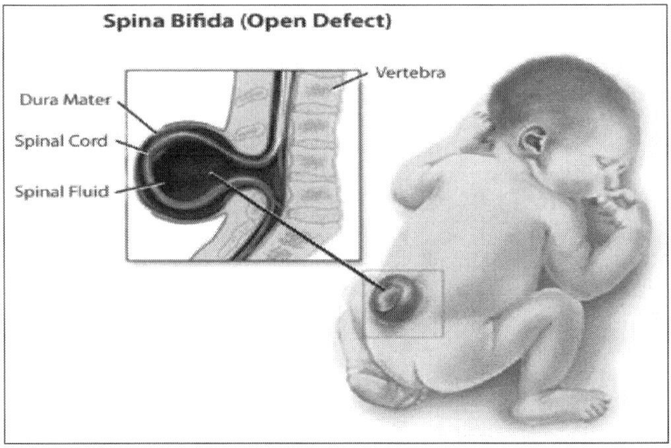

Requirements/RDA of Folic Acid

Age Group	*Requirements (mg)*
Infant	25
Preschool children	80
Adolescents	200
Adult	200
Pregnancy	500
Lactation	300

Pyridoxine (B_6)

Pyridoxine is the general term for group of vitamins has similar functions. Three forms of vitamin B6 occur in nature-Pyridoxine, Pyridoxal and Pyridoxamine. Pyridoxinplays major role in protein metabolism. It is heat stable, soluble in water, sensitive to light and alkali.

Functions

1. Pyridoxine plays major role in protein metabolism.
2. It is essential for the growth of infant.
3. It aids in the formation of synthesis of haemoglobin.
4. Pyridoxine is essential for deamination and decarboxylation.
5. Vitamin B_6 is involved in the conversion of tryptophan into niacin.
6. It aids in the conversion of linoleic acid to arachidonic acid.
   It act as coenzyme in carbohydrate and fat metabolism
7. It aids in the release of glycogen from liver and muscle.

Food Sources

Meat, pulses and wheat

Effects of Deficiency

Microcytic anaemia, abnormal protein metabolism, convulsions, nervousness, irritability, insomnia.

Requirements/RDA of Vitamin B_6

Age Group	*Requirements (mg)*
Infant	0.3-0.4
Preschool children	0.6
Adolescents	2.0
Adult	2.0
Pregnancy	2.5
Lactation	2.5

Vitamin B_{12} (Cyanocobalamine)

Pharmaceutical form of B12 is cyanocobalamine. Isolated in 1948 from liver. The name cyanocobalamine is due to the presence of cobalt and cyanide (CN) group in their structure. Other name of Vitamin B12 is antipernicious anaemia factor and extrinsic factor of castle. It contains **cobalt**. It is sensitive to sun light.

Functions

1. Vitamin B_{12} is necessary for normal growth, healthy nervous tissue and normal blood formation.
2. Vitamin B_{12} is involved in DNA synthesis.
3. It is essential for the formation of red blood cells.
4. It facilitates the formation of folate co-enzymes needed for nucleic acid synthesis.
5. Vitamin B_{12} is also required for the synthesis of myelin sheath that surrounds the nerve fiber.

Food Sources

Vitamin B_{12} is present **only in foods of animal origin**. Liver sheep, shrimp, mutton, egg, milk are good sources of Vitamin B_{12}.Vitamin $B_{12}$is synthesized by the colonic bacteria.

Deficiency Disease

Pernicious anaemia (immature red blood cells with normal amounts of haemoglobin), soreness and inflammation of tongue, numbness and tingling in fingers and toes.

Requirements/RDA of Vitamin B_{12}

Age Group	*Requirements (mg)*
Infant	0.2
Preschool children	0.2-1.0
Adolescents	1.0
Adult	1.0
Pregnancy	1.2
Lactation	1.5

Biotin

In 1927, Boas made an important observation that when raw egg white was incorporated as then main source of protein in the diets of rats, they developed deficiency symptoms characterized by dermatitis, loss of hair and muscular in coordination. All these symptoms were prevented or cured by yeast, liver and egg yolk.

Biotin is sparingly soluble in cold water freely soluble in hot water. Sparingly soluble in alcohol but insoluble in fat.

Physiological Functions

1. Biotin is essential for normal gestation and lactation in experimental animals.
2. It helps to maintain the skin and the nervous system in sound conditions.
3. It is essential for the synthesis of malonyl COA from acetyl COA and oxalaoacetic acid from pyruvic acid.

Effects of Deficiency

Maculosquamous dermatitis, brawny desquamation, depression, extreme lassitude, somnolence, muscle pain, and hyperaesthesia.

Egg White Injury Factor (Avidin)

Avidin is the active principle present in egg white and responsible for producing 'egg white injury' in experimental animals. **Avidin** present in egg white which **binds biotin** and make it into unavailable form.

Avidin present in egg white inactivated by prolonged heating (steaming at 100°C for 60 minutes).

Dietary Sources

Dried yeast, rice polishing, wheat germ, liver, peanut, soya bean

Requirements

The minimum daily requirements for biotin are not known.

Pantothenic Acid

Pantothenic acid is one of the vitamins of the vitamin B complex which can prevent or cure a specific type of dermatitis. Pantothenic acid is the constituents of Coenzyme A present in all living mater. In 1939 R.J. Williams isolated the vitamin and called it as pantothenic acid. CO- enzyme form of pantothenic acid is coenzyme A

Physiological Functions

- Pantothenic acid in the form of coenzyme A takes part in the metabolisms of carbohydrates and fats.
- CO enzyme A is essential for synthesis of fatty acids and their incorporation in to membrane phospholipid.
- Essential for synthesis of cholesterol, steroid hormones, Vit A and D.

- It is also contribute to synthesis of amino acids, *e.g.*, Leucine, arginine, methionine
- It is essential for the oxidation of pyruvic acid.

Effects of Deficiency

- Nausea, vomiting, tremor of the outstretched hand and irritability
- Burning feet syndrome.

Chapter 9

Minerals: Functions, Sources, Requirements and Deficiency Diseases

Minerals are elements that originate in the Earth and cannot be made by living organisms. They are, therefore, inorganic do not contain carbon molecule. The Plants obtain minerals from the soil, and most of the minerals in our diets come directly from plants, however this varies with geographic locale because of soil mineral content. Minerals are sturdy nutrients, meaning they will not be destroyed during most food preparation methods, although some leach out into the water used in various cooking methods.

About 17 minerals are considered 'essential' to humans required for proper functioning and good health. Out of these 17 which are required more than 100mg per day is called as macro minerals which are required less than 20 mg per day is called as micro minerals or trace elements

Macro minerals - Calcium, Phosphorus, Sodium, Magnesium, and Potassium

Micro minerals/Trace element - Iron, Iodine, Zinc, Fluorine, Chlorine

Calcium

Calcium is an essential element required for several life processes. Most abundantly present in human body. The function and requirement of Calcium and Phosphorous are closely linked. Over 99 per cent of the Calcium and Phosphorous is present in the bones and the remaining 1 per cent in the body fluids. The Calcium and Phosphorous are present in the ratio of **2:1 in our body**. Adult body contains about 1000 to 1200 g of calcium. The **diet** should have calcium and phosphorous in ratio of **1:2.** Vitamin D is essential for the absorption calcium. Blood Ca level is 9-11 mg/dl and it is maintained by the two hormones they are **calcitrol** and **parathyroid hormone-aided by calcitonin.**

Functions

1. It is essential for the formation of bone.
2. It is essential for the formation of teeth.
3. Calcium is essential for the clotting of blood.
4. It is essential for the muscle contraction and relaxation
5. It is essential for nerve impulse transmission
6. Essential for cell membrane permeability
7. It is essential for Activation enzymes
8. Essential for release of hormones
9. Essential for action of Muscle (heart muscles)

Food Sources

Ragi, Milk and milk products, bengal gram, gingelly seeds, cumin seeds, agathi, amaranth, drumstick,

Deficiency Diseases

- ✰ Osteoporosis (decalcification of bone leads to brittle bone)
- ✰ Osteomalacia
- ✰ Tetany

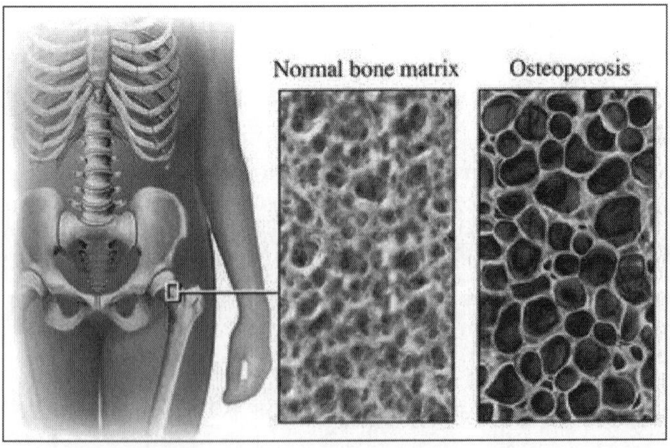

Requirements/RDA of Calcium

Age Group	*Requirements (mg)*
Infant	500
Preschool children	600
Adolescents	800
Adult	600
Pregnancy	1200
Lactation	1200

Phosphorus

Phosphorus is a essential mineral in every cells in the human body. Adult body contains about 400-700 g (approximately 1 per cent of the body weight) of phosphorus. Majority of the phosphorus is occurred in the form of Phosphate (PO_4). The Calcium and Phosphorus are present in the ratio of 2:1 in our body. About 85 per cent of the phosphorus is in bone and teeth, Serum phosphorus level is level 2.5-5mg/dl and is higher in children's as 4 – 7 mg/dl

Functions

1. Essential for Bone and teeth formation
2. Essential for facilitation of energy transaction
3. Essential for transport of nutrients
4. Essential for of protein activity
5. Components of essential body components
6. Essential for Regulation of acid base balance

Food Sources

Ragi, Milk and milk products, egg, meat, bengal gram, gingelly seeds, cumin seeds, agathi, amaranth, drumstick.

Deficiency Diseases

- Myopathy
- Growth failure
- Defects of leucocyte function

Requirements/RDA of Phosphorus

- Infant- 500 mg
- Adult – 1200 mg

Magnesium

The adult body contains about 20 to 25 mg of magnesium. It is present in the bones in combination with phosphate and carbonate. 60-70 per cent of the magnesium present in the bone, 25-30 per cent in muscle, 6-8 per cent soft tissues.

Functions

1. It acts as activator of several enzymes, *e.g.*, alkaline phosphatase.
2. It is required as a co-factor for oxidative phosphorilation.
3. Magnesium is present in certain enzyme, *e.g.*, co-carboxylase

Effect of Deficiency

Depression, muscular weakness, convulsion

Source

Cereals, pulses, almond, cashew nut, peanut, milk, pork *etc.*

Requirements/RDA of Magnesium

Age Group	*Requirements (mg)*
Infant	30-45
Preschool children	50
Adolescents	235
Adult	310-340
Pregnancy	310
Lactation	310

Potassium

The human adult body contains about 250 g of potassium. Large portion of potassium is present inside the cell. It act as a major cation in the intracellular fluid, a Small amount in extra cellular fluid-significantly effect on muscle activity – heart muscle. Blood concentration of Potassium is 14-20 mg/dl. All natural foods are rich source of potassium.

Functions

1. It plays important role in the regulation of the acid base balance in the cells.
2. It is essential for the muscular contraction.
3. It is essential for the growth and builds up of tissues as it is required for the synthesis of protein.
4. Essential for maintenance of water balance

5. Essential for maintenance of blood pressure
6. Essential for the release of insulin from pancreas
7. It is essential for metabolic function

Source

Almond, cashew, peanut, pulse, fruits, meat, fish are rich source of potassium.

Deficiency

Less intake of food during starvation and surgery is responsible for the deficiency of potassium. Excessive loss might occur during vomiting and diarrhea for long duration. In severe burns and prolonged fever potassium deficiency occurs. Patient might complains of nausea, vomiting, muscle weakness, low blood pressure and tachycardia.

Hyperpotassiumia

Excess of potassium is frequent complication in renal failure. The symptoms include muscular weakness, poor respiratory system which later on leads to cardiac failure.

Sodium

Sodium chloride is the only mineral which is taken in more or less pure form in addition to the amount present in natural foods. It is the major cation of extracellular fluid and it is present 120 mg in adult body of which 1/3 is bound in skeletal system- 40 per cent, 2/3 is in extracellular fluid compartment – 50 per cent and the 10 per cent or less in intracellular fluid. Approximately. Sodium chloride or common salt is the taste-giver in our diet.

Functions

1. It regulates acid-base balance of the body.
2. It regulates the osmotic pressure of plasma.
3. It regulates the fluid balance in the body
4. Na ions play an important part in the absorption of monosaccharide and amino acids from small intestine.
5. Na ions play an important part in maintaining the heartbeat.
6. It regulates the normal Muscle irritability

Effect of Deficiency

Intense painful contraction of skeletal muscle is called heat cramps. It is caused due to the excessive NaCl losses from body by sweating.

Low Sodium Diet

Low sodium diet (200-400 mg Na or NaCl) is prescribed for patients suffering from hypertension.

Requirements/RDA of Sodium

Age Group	*Tropical (g)*	*Temperate (g)*
Adult: light work	10	5
Moderate work	15-20	8
Heavy work	25-30	15
Children	5	3-5
Adolescents	10-25	6-8

Food Sources

Cereal products, cashew nuts, ground nut, milk, cured pork, salted fish

MICRO MINERALS /TRACE MINERALS

Important micro minerals are Iron, Zinc, Copper, Sodium, Cobalt, Fluoride, Manganese, Chromium, Iodine and Molybdenum.

Iron

The total body iron is 4-5 g in adults. Iron present as haemoglobin in our body. It is essential for the formation of blood. About 70 per cent of iron is in circulating hemoglobin, 4 per cent into myoglobin. 25 per cent is stored in liver, spleen, kidney and 1 per cent in plasma and various oxidative enzymes.

Functions

- ✰ Co factor of enzymes and other protein
- ✰ Formation of red blood cells

1. Iron essential for the formation of red blood cell- haemoglobin.
2. It is essential for transport of oxygen and storage of oxygen in the muscle by combining with hemoglobin
3. It forms a part of the myoglobin in muscles which makes oxygen available for muscle contraction.
4. Act as a cofactor of several enzymes and other protein. As part of enzymes iron catalyzes many important reactions in the body.

Examples are

a) Conversion of beta carotene to active form of Vitamin A.

b) Synthesis of collagen and neurotransmitters.

c) Detoxification of drugs in the liver.

d) It is essential for the transport and storage of oxygen.

e) Synthesis of Carnitine- a vitamin like substance essential for the transport of fatty acids

f) Synthesis of purines, which form an integral part of DNA, RNA

g) synthesis of neurotransmitter topamine, serotonin and norepinephrine

Food Sources

Ragi, green leafy vegetables, dried fruits, jaggery, liver, fish, poultry, meat, eggs, dates *etc.*

Deficiency Diseases

Nutritional Anaemia

Anaemia means a deficiency of red blood cells. Anaemia occurs when the haemoglobin level falls below 12 gm/dl in adult man and woman. During pregnancy haemoglobin level below 11 gm/dl is termed anaemia.

This is caused by the dietary lack of iron or inadequate absorption and utilization of iron. Nutritional anaemia is the common form of anaemia affecting women in reproductive years, infants and children.

Causes of Anaemia

1. Delayed weaning of infant may causes anaemia
2. In females menstruation causes an average loss of 30 mg of iron each month. So it requires approximately 1 mg a day in addition to the normal needs. Inadequate intake may causes anaemia,
3. Inadequate iron intake in diet
4. Loss of iron or malaborption of iron.
5. Reduction in gastric acid output in gastritis may be cause of anaemia.
6. Ulcers, parasitic infestations may also be gastritis may be a cause of anaemia.

Signs and Symptoms of Nutritional Anaemia

1. Reduced Haemoglobin level (less than 12 g/dl)
2. Paleness of skin
3. Inside of the lower eyelid is pale pink
4. Finger nails becoming thin and flat (spoon shaped nails). Flattening of nails called **koilonychia**
5. Fatigue, breathlessness on exertion
6. Sleeplessness, dimness of vision

7. Weakness, fatigue and headache
8. Loss of appetite
9. Burning sensation in the stomach
10. Insomnia
11. Women or children start eating strange things such as coal, soil, chalk *etc.* This is called **PICA**
12. Liver and spleen become enlarge

Types of Anaemia

1. **Hypochromic anaemia:** In this type of anaemia red blood cells are paler (less redness) than normal.
2. **Sideroblastic Anaemia:** In this condition red blood cell production is impaired by disordered iron metabolism. Heredity can be one of cause.
3. **Megaloblastic anaemia:** It is a condition in which bone marrow produces unusually large, structurally abnormal, immature red blood cell. It is results from inhibition of DNA synthesis during red blood cell production.
4. **Pernicious anaemia:** It is also known as vitamin B_{12} deficiency anaemia. It is a condition in which there are not enough red blood cells due to lack of vitamin $B_{12.}$

Excess Intake iron

If iron intake is excess in amount, it will deposit in liver. This condition is known as **Hemosiderosis**. It is due to abnormal metabolism of iron.

Requirements/RDA of Iron

Age group	Requirements (mg)
Adult: Men	17
Women	21
Pregnant women	35
Lactating women	21
Adolescents: Boys	28
Girls	26

Iodine

Iodine is an essential constituent of the thyroid hormone produced by the thyroid glands. It occurs as free iodide ions or as protein bound iodine in our body. About 15 – 23 mg of iodine is present in the adult human body.

Function

Normal growth and development

Iodine is essential for the synthesis of the thyroid hormones T3 and T4.

Sources

Richest source of iodine are sea foods like sea fishes and common salt from sea water.

Deficiency Diseases

Goitre

The enlargement of thyroid glands due to the inadequate of iodine is called goitre.

Particulars	*Weight of Thyroid Glands (g)*
Normal condition	25
Goitre	200-500

Cretinism

The retardation in physical and mental growth of children due to severe iodine deficiency is known as cretinism.

The symptoms of cretinism are:

1. Low basal metabolic rate
2. Muscular flabbiness and weakness
3. Dry skin
4. Rough hair
5. Enlarge tongue
6. Thick lips retardation of skeletal development
7. Mental retardation

Hypothyroidism

It is characterised by dry skin, husky voice, delayed tendon reflexes.

Myxedema

Iodine deficiency in adults results in Myxedema. Face becomes thick and looks puffy. Facial expression is poor and person becomes inactive.

Requirements/RDA of Iodine

Age Group	*Requirements (μg)*
Adult	100-200
Pregnant women	+25
Lactating women	+50
Infant	40-50

Goitrogens

Goitrogens are substances present in foods which cause goitre. These substances react with iodine present in the food and making it unavailable for absorption. Foods like cabbage, cauliflower, radish contain goitrogens.

Zinc

Its total quantity in the body is 1.4-2.3 g. The largest stores of Zinc are present in the bones. Zinc forms a constituent of the blood.

Functions

1. Zinc is essential for the brain development of children
2. It is essential for the spermatogenesis and development of primary and secondary sex organs
3. It is essential for quick wound healing
4. It is essential for normal taste sensation
5. Zinc is a constituent of enzymes such as carbonic anhydrase, alkaline phosphatase, and lactic dehydrogenase
6. It is a constituent of the hormone insulin
7. It plays a major role in the synthesis of DNA and proteins

Sources

Meat, unmilled cereals, legumes, nuts and oil seeds are good sources.

Deficiency Diseases

Acrodermatitis enteropathica (severe dermatitis in infants), emotional disorder, impaired taste (hypogeusia) and smell, delayed genital development, immunological defects, impaired glucose tolerance.

Requirements/RDA of Zinc

Age Group	*Requirements (mg)*
Adult: Men	12
Women	10
Pregnant women	12
Lactating women	12
Adolescence: Boy	12
Girls	12

Fluorine

Fluorine occurs in the form of fluoride in nature. It is present in soils, water, plants and animals.

Functions

1. Fluorine in the teeth helps to protect them from decay
2. Fluoride gives stability to bone and enamel tissues

Deficiency

- ✰ Tooth decay
- ✰ Osteoporosis

Toxic Effects of Excess of Fluorine

1. Skeletal Fluorosis

Intake of fluorine higher than the 10 mg/litre of drinking water leads to skeletal fluorosis (genuvalgum- knock knees). In skeletal fluorosis there is hypercalcification of the bone of the spine, pelvis, and limbs.

2. Dental Fluorosis (Mottled Enamel)

Intake of fluorine higher than the 3-5 mg/litre of drinking water leads to dental fluorosis. Chalky white patches, brown staining are the symptoms of dental fluorosis.

Source

Soil and water are the major source of fluorine.

Requirements/RDA of Fluorine

Age Group	*Requirements (mg)*
Adult	1.5 – 4.0
Adolescents	1.5-2.5
Infant	0.1-1.0

Chlorine

Chlorine exists in the form of a salt (chloride) in the human body. It is present mostly in the extra cellular fluid. In the body it is mostly present as sodium chloride or common salt, which performs various essential functions. Combining with sodium, it forms the most important electrolyte in the body.

Functions

1. The osmotic pressure in the cells and tissues is well maintained when there is optimal chloride level in the blood.
2. Chloride is a catalyst in manufacturing of hormones from different glands.
3. It also prevents build up excessive fat in tissues.
4. Chloride being an electrolyte helps and maintains proper functioning of muscles. With loss of electrolytes the muscles become tensed and contracted. They lose their efficiency in performing their function. As chlorides are lost in sweat, people doing heavy exercise tend to lose it and thus manifest problems related to chloride and other electrolyte deficiencies.
5. The acid and base balance in blood is regulated with chloride.
6. It combines with potassium to form potassium chloride. This component is required for production of hydrochloric acid in stomach. The acid is used for breaking down food in stomach into various nutrients that are ultimately absorbed further down in the small intestine. Especially for digestion of protein.
7. Adequate chloride in body helps to maintain proper fluid and electrolyte balance in the body.

Deficiency of Chloride

- One of the reasons for deficiency of chloride is when salt (sodium chloride) is restricted in patients suffering from swelling in leg and different parts of body (edema). It is also restricted in patients suffering from high blood pressure. In such cases, the symptoms produced are same as that of sodium chloride deficiency. The skin elasticity is lost. Tongue becomes dry.

- Excessive loss of this salt in perspiration and diarrhea can lead to muscular cramps.
- Chloride deficiency can lead to loss of teeth and hair.
- Food digestion may be affected due to deficiency of chloride as it is involved in manufacturing of hydrochloric acid.
- Fluid levels may get deranged.

Sources

Chloride is found in barley, wheat, and other grains and pulses, green leafy vegetables, and fruits like melon, and pineapple

Requirements/RDA for Chloride

Age Group	*Requirements (g/day)*
Infants – Light work	10-15
Infants – Hard work	15-20
Infants – Very hard work	25-30
Children	5-10
Adolescent boys and girls	10-25
Women, Pregnancy: first half	10
Womcn, prcgnancy: 2nd half	5
Lactation	15

Chapter

10

Water: Importance of Water and Maintenance of Electrolyte Balance, Health Benefits of Fibre

What is Water?

Water is one of the most **important** substances on earth. All plants and animals must have **water** to survive. If there was no **water** there would be no life on earth. Apart from drinking it to survive, people have many other uses for **water**.

With two thirds of the earth's surface covered by water and the human body consisting of 60 to 75 percent of it, it is evidently clear that water is one of the prime elements responsible for life on earth. We can live without food for extended periods of time, but without water will result in death. Water is colourless, calorie less compound of hydrogen and oxygen that virtually every cell in the body needs to survive.

An average adult body contains 42 litres of water and with just a small loss of 2.7 litres he or she can suffer from dehydration, displaying symptoms of irritability, fatigue, nervousness, dizziness, weakness, and headaches.

Importance of Water

- ☆ Essential constituent of all the cells of the body
- ☆ Transporting nutrients and removes waste products
- ☆ Medium for most biochemical reactions

- ☆ Solvent for electrolytes, non electrolytes, hormones, enzymes and vitamins
- ☆ Maintenance of body temperature
- ☆ Constituents of several enzymes and hormones. Saliva is about 99.5 percent water
- ☆ Maintaining the form and texture of the tissues
- ☆ Maintenance of acid base and electrolyte balance
- ☆ Lubricant

Distribution of water

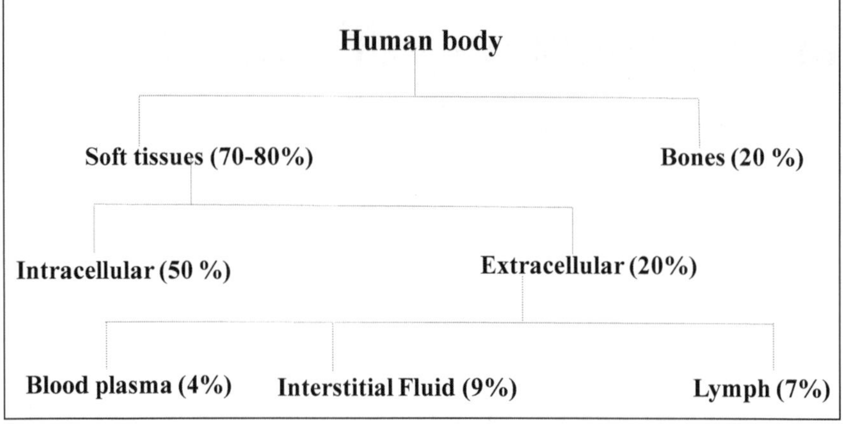

Water Intake

Water Input	*Water Intake (ml)*	
	Temperate	*Tropical*
Drinking water	1500	2000-5000
Food	1000	1000-2000
Oxidation-Carbohydrate, protein and fat	300	300
Total	2800	3300-7300

Water Loss from Body

Excretion	*Water Loss (ml)*	
	Temperate Climate	*Tropical Climate*
Kidney	1500	1000-1500
Skin	800	1800-5200
Lungs	400	400
Faeces	100	100-200
Total	**2800**	**3300-7300**

Maintenance of Electrolyte Balance

Electrolytes are minerals in your body that have an electric charge. They are in your blood, urine and body fluids. Maintaining the right **balance** of **electrolytes** helps your body's blood chemistry, muscle action and other processes. Sodium, calcium, potassium, chlorine, phosphate and magnesium are all **electrolytes**. Levels of **electrolytes** in your body can become too low or too high. That can happen when the amount of water in your body changes, causing dehydration or over hydration.

Electrolytes are the smallest of chemicals that are important for the cells in the body to function and allow the body to work. Electrolytes such as sodium, potassium, and others are critical in allowing cells to generate energy, maintain the stability of their walls, and to function in general. They generate electricity, contract muscles, move water and fluids within the body, and participate in myriad other activities.

Electrolyte Balance

The concentration of electrolytes in the body is controlled by a variety of hormones, most of which are manufactured in the kidney and the adrenal glands. Sensors in specialized kidney cells monitor the amount of sodium, potassium, and water in the bloodstream. The body functions in a very narrow range of normal, and it is hormones like renin (made in the kidney), angiotensin (from the lung, brain and heart), aldosterone (from the adrenal gland), and antidiuretic hormone (from the pituitary) that keep the electrolyte balance within those normal limits.

To function normally, the body must keep fluid levels from varying too much in the areas of the body that contain fluid (called compartments). The three main compartments are

- Fluid within cells
- Fluid in the space around cells
- Blood

Electrolytes, particularly sodium, help the body maintain normal fluid levels in these compartments (called fluid balance) because the amount of fluid a compartment contains depends on the concentration of electrolytes in it. If the electrolyte concentration is high, fluid moves into that compartment. If the electrolyte concentration is low, fluid moves out of that compartment. To adjust fluid levels, the body can actively move electrolytes in or out of cells. Thus, having electrolytes in the right concentrations (called electrolyte balance) is important in maintaining fluid balance among the compartments.

The kidneys help maintain electrolyte concentrations by filtering electrolytes from blood, returning some electrolytes to the blood, and excreting any excess into the urine. Thus, the kidneys help maintain a balance between daily consumption and excretion.

If the balance of electrolytes is disturbed, disorders can develop. An electrolyte imbalance can result from the following

- ☆ Becoming dehydrated
- ☆ Taking certain drugs
- ☆ Having certain heart, kidney, or liver disorders
- ☆ Being given intravenous fluids or feedings in inappropriate amounts

Dietary Fiber

Dietary fiber is defined as that portion of plant material ingested in the diet that is resistant to digestion by gastro intestinal secretions. It consists of hemicellulose, cellulose, lignins, oligosaccharides, pectins, gums and waxes.

Fiber, also known as roughage, is the part of plant-based foods (grains, fruits, vegetables, nuts, and beans) that the body can't break down. It passes through the body undigested, keeping your digestive system clean and healthy, easing bowel movements, and flushing cholesterol and harmful carcinogens out of the body.

Fiber comes in two varieties: insoluble and soluble.

- ☆ Insoluble fiber does not dissolve in water. It is the bulky fiber that helps to prevent constipation, and is found in whole grains, wheat cereals, and vegetables such as carrots, celery, and tomatoes.
- ☆ Soluble fiber dissolves in water and helps control blood sugar levels and reduce cholesterol. Good sources include barley, oatmeal, beans, nuts, and fruits such as apples, berries, citrus fruits, and pears

Many foods contain both soluble and insoluble fiber. In general, the more natural and unprocessed the food, the higher it is in fiber. There is no fiber in meat, dairy, or sugar. Refined or "white" foods, such as white bread, white rice, and pastries, have had all or most of their fiber removed.

Health Benefits

- ☆ **Digestive health** - Dietary fiber normalizes bowel movements by bulking up stools and making them easier to pass. This can help relieve and prevent both constipation and diarrhoea. Eating plenty of fiber can also reduce your risk for diverticulitis (inflammation of the intestine), hemorrhoids, gallstones, kidney stones, and provide some relief for irritable bowel syndrome (IBS). Some studies have also indicated that a high-fiber diet may help to lower gastric acid and reduce your risk for gastro esophageal reflux disorder (GERD) and ulcers.
- ☆ **Heart disease** - Fiber, particularly soluble fiber, is an important element of any heart-healthy diet. Eating a diet high in fiber can improve cholesterol levels by lowering LDL (bad) cholesterol. A high fiber intake can also reduce your risk for metabolic syndrome, a group of risk factors linked

to coronary heart disease, diabetes, and stroke. Fiber can help to lower blood pressure, reduce inflammation, improve levels of HDL (good) cholesterol, and shed excess weight around the abdomen.

- **Diabetes -** A diet high in fiber particularly insoluble fiber from cereals can lower your risk for type 2 diabetes. If you already have diabetes, eating soluble fiber can slow the absorption of sugar and improve your blood sugar levels.
- **Cancer -** There is some research that suggests eating a high-fiber diet can help prevent colorectal cancer, although the evidence is not yet conclusive. Diets rich in high-fiber foods are also linked to a lower risk for other common digestive system cancers, including stomach, mouth, and pharynx.
- **Skin health** - When yeast and fungus are excreted through the skin, they can trigger outbreaks or acne. Eating fiber, especially psyllium husk (a type of plant seed), can flush toxins out of your body, improving the health and appearance of your skin.

A high-fiber diet has many benefits, which include:

- Normalizes bowel movements
- Dietary fiber increases the weight and size of your stool and softens it
- Helps maintain bowel health
- Lowers cholesterol levels
- Helps control blood sugar levels
- Aids in achieving healthy weight

RDA for Dietary Fiber

Recommended dietary allowances for fiber is 30 g/day or 12g/1000 Kcal for normal healthy adult. RDA for dietary fiber is 25-38 g of per day per persons suffering from diabetes.

Chapter 11

Preservation of Food by Low and High Temperature and Food Irradiation

Preservation

Preservation are the measures employed to prevent (or) minimize all such undesirable changes takes place due to spoilage which caused by both internal and external organisms.

Need for Preservation of Foods

Freshly harvested or prepared products are highly attractive in appearance and possess good taste and aroma, but deteriorate rapidly if kept for sometimes. This is due to (1) Fermentation caused by moulds, yeasts and bacteria (2) Enzymes present in the products may affect the colour and flavour adversely (3) Chemicals present in the processed food react with one another and spoil its taste and aroma (4) Air coming in contact with the product, may react with the glucosidal materials present in it and render the product bitter (5) Traces of metal from the equipment may get into the product and spoil its taste and aroma.

Importance/Advantages of Food Preservation

1. Prevention of wastage of fresh produce
2. To extent the shelf life of the produce
3. To develop value added products

4. Self-employment opportunities
5. Provides convenient and preferable forms to the consumer (jam, jellies, RTS, squash *etc.*)
6. Foreign exchange earnings from the export of processed foods and it will improve our national economy
7. Available in the form of ready to serve and hence (fuel consumption) energy is saved

In the preservation of foods by various methods, the following principles are involved:

1. Prevention or Delay of Microbial Decomposition

- By keeping out microorganisms (asepsis)
- By removal of microorganisms, *e.g.*, by filtration
- By hindering the growth and activity of microorganisms, *e.g.*, by low temperature, drying, anaerobic conditions, chemicals or antibiotics; and
- By killing the microorganisms, *e.g.*, by heat or radiation

2. Prevention or Delay of Self-Decomposition of the Food

- By destruction or inactivation of enzymes, *e.g.*, by blanching
- By prevention or delay of chemical reactions, *e.g.*, prevention of oxidation by means of an antioxidant

3. Prevention of Damage by Insects, Animals, Mechanical Causes *etc.*

To retain the natural taste and aroma of a product, it is necessary to preserve it soon after preparation, without allowing it to stand for any length of time.

Various methods of preservation are employed and each has its own merits. The methods generally used areas under:

Asepsis (Absence of infection)

Asepsis means preventing the entry of microorganisms. Proper packaging of products prevents the entry of microorganism.

Maintaining of general cleanliness while, picking, grading, packing and transporting of fruits and vegetables increases their keeping quality

Washing or wiping of the fruits and vegetables before processing reduces the number of microorganism.

Preservation by High Temperature

Coagulation of proteins and inactivation of their metabolic enzymes by the application of heat leads to the destruction of microorganisms present in foods.

Further, heating can also inactivate the enzymes present in the food. The following methods are example for the preservation of food by high temperature.

- Pasteurisation
- Sterilisation
- Aseptic canning
- Hot pack or hot fill

Pastuerization

Heating Food Below 100°C

- Pasteurization does not kill all microorganisms. Some spores and spore forming bacteria can survive and multiple later.
- TDT is defined as the time required at a given temperature to kill a stated number of organisms under specified conditions.

  e.g. Salmonella typhosa 4.3 mts at 60°C

  E. coli 20-30mts at 57.3°C

  L. bulgaricus 30mts at 71°C

Types of Pastuerization

1. Bottle or Holding Pasteurization

- After filling the juice by leaving the sufficient head space the bottles are sealed and pasteurized.
- The head space is to allow for expansion of juice during heating
- Commonly used for the preservation of fruit juices at home.

2. Overflow Method

- Juice is heated to a temperature about 2.5°C higher than the pasteurization temp, then fill in the bottles, at pasteurization temp and after sealing the second pasteurization is done at 2.5°C lower than filling and sealing temperature.

3. Flash Pasteurization

- The juice is heated rapidly to temperature of about 5.5°C higher than the pasteurization temperature for about a minute.
- The method has been developed specially for the canning of natural orange juice but can also be used for grape and apple juices.

Advantages

1. Loss of flavour is minimum

2. Vitamins are not destroyed
3. Effects economy of time and space
4. Keeps the juice uniformly cloudy, and
5. Juice is heated uniformly and thus its cooked taste is minimum

Sterilization

- Sterilization means the destruction of all viable microorganisms. Heat sterilization is the most effective process of food preservation. By this method all microorganisms are completely destroyed due to high temperature. The time and temperature necessary for sterilization vary with the type of food.
- Fruits -100°C for 30 minutes
- Vegetables-116°C for 30 to 90 minutes
- Temperatures above 100°C can only be obtained by using steam pressure sterilizers such as pressure cookers and autoclaves

Difference Between Pasteurization and Sterilization

Pasteurization	*Sterilization*
1. Partial destruction of microorganisms	1. Complete destruction of micro organisms
2. Temp below 100°C	2. Temp 100°C and above normally used for vegetable
3. Normally used for fruits	

Aseptic Canning

- Is a quick technique in which food is sterilized outside the can and aseptically placed in previously sterilized cans, which are sealed in aseptic environment.
- This is known as martin aseptic canning

Hot Pack or Hot Fill

- Hot pack or Hot fill - used is houses for preparation of Jam and other products
- Hot pack or Hot fill refers to the filling of previously pasteurized or sterilized food while still hot in to clean but not necessarily sterile conditions.

Preservation by Low Temperature

Microbial growth and enzyme reactions are retarded when foods are stored at low temperatures. Lower the temperature, the greater the retardation. Low temperatures can be produced by

I. Cellar storage (about 15°C)
II. Refrigeration or chilling (0 to 5°C)
III. Freezing (-18 to -40°C)

I. Cellar Storage (Cellar - underground rooms)

- The temperature in cellar where surplus food is stored in many villages is usually not much below that of the outside air but lower than 15°C.
- It is low enough to prevent the action of many spoilage organisms (or) of plant enzymes.

II. Refrigeration or Chilling (0 to 5°C)

- Chilling temperature is obtained and maintained by means of ice or mechanical refrigeration.
- Fruits and vegetables and their products can be preserved for few days to many weeks when kept at this temperature. The best storage temperature for many foods is slightly above 0°C but varies with products.

III. Freezing (-18 to -40°C)

- Freezing method is the most harmless method of food preservation. Microbial growth is inhibited and the rate of chemical reactions is slowed down at low temperatures.
- At temperatures below the freezing point of water growth of microorganisms and enzyme activity are reduced to a minimum. Most perishable food scan be preserved for' several months.

(1) Sharp Freezing (Slow freezing)

- The temperature may vary from -15 to -29°C.
- Freezing may take from 3 to 72 hours.
- Large ice crystals formed and rupture the cells. The thawed tissue cannot regain its original water content.
- The first products to be sharp frozen were meat and butter.

(2) Quick Freezing

- Freezing may takes within 30 minutes or less.
- Very small ice crystals are forms

- It will not of rupture the cells
- Most foods are quick frozen by one of the following three methods:

(A) By Indirect Contact with Refrigerant

- Indirect freezing may be defined as freezing by contact of the product with a metal surface.
- The metal cooled by freezing brine or other refrigerating media.
- This is an old method of freezing in which the food or package is kept in contact with the passage through which the refrigerant at -18 to -46°C.

Types of Indirect Contact Freezing

- Plate freezing system
- Indirect-contact freezing system
- Cabinet freezing system
- Scraped surface freezing system

Plate Freezing

- Packaged or fresh commodities are placed over the surface of plate cooled by refrigerant in a cylindrical scraped surface heat exchanges.
- Plates are the primary barrier between the cold refrigerant and the product
- Freezers consist of a vertical or horizontal series of hollow plates
- Refrigerant is pumped at -40°C temperature
- Product is pressed between hollow metal plates, either horizontally or vertically, with a refrigerant circulating inside the plates
  - Vertical plate freezers are suitable for the viscous products
  - Horizontal plate freezers are suitable for packaged products
- Flat and relatively thin foods are placed in single layers between plates, slight pressure is applied by moving plates together which improves contact between the food and plates

Cabinet Freezing

- Product placed in a package prior to freezing
- Packages are placed on trays which are moved into the freezing system
- Operate as batch systems, with freezing time established by the length of time that the product remains in the cabinet
- Environment in the room need to be with low temperature, and air movement is established by fans within the cabinet

Indirect-Contact Freezing System

- Use high-velocity air
- Product package is the barrier between cold air and product within the package
- Continuous systems
- Continuous product movement in a direction concurrent to air flow

Scraped Surface Freezing

- Liquid/semi-solid foods - ice cream
- Utilize a scraped surface heat exchanger – primary component of the continuous system
- Used to convert liquid product into a frozen slurry
- A jacket (freezer barrel) surrounding a high speed rotor, fitted with short blade with a outer wall of the heat exchanger barrel represents the barrier between the product and the low-temperature refrigerant
- Rotor scrapes frozen food from the wall of the freezer barrel and incorporates air
- Freezing is very fast, 50 per cent of the water is frozen within a few seconds
- Very small crystals formed
- Smooth creamy consistency to the product
- Temperature is reduced between -4°C and -7°C
- Frozen aerated mixture is then pumped into containers
- Freezing is completed in hardening room

(B) Direct Contact between the Product and Medium

- Used for reduction of product temperature
- Appropriate and effective
- No barrier between product surface and the medium

Direct Contact Freezing Systems

- Air blast system
- Spiral conveyor system
- Fluidized bed system
- Immersion freezing system

By Air Blast (Air blast freezing)

- This refers to vigorous circulation of cold air in order to freeze the product.

- ☆ Freezing is done by placing the foodstuffs on trays or on a belt which are then passed slowly through an insulated tunnel containing air in it.
- ☆ Here the air temperature is approximately -18 to - 34°C or even lower. Vegetables are frozen by blowing cold air.

Direct Immersion Method

- ☆ Food can be frozen rapidly by direct immersion in a liquid such as-brine 'or sugar solution at low temperature.
- ☆ The refrigeration medium must be edible and capable of remaining unfrozen at -18°C and slightly below.

Advantages

- ☆ There is perfect contact between the refrigerating medium and the product.
- ☆ Fruits are frozen with a coating of syrup which preserves the colour and flavour during storage.

Disadvantages

- ☆ Brine is a good refrigerating medium but it cannot be used for fruits.
- ☆ It is difficult to make a syrup that will not become viscous at low temperature

(3) Cryogenic Freezing

- ☆ Mushrooms, sliced tomatoes, whole strawberries and raspberries are frozen by this method.
- ☆ They are frozen at very low temperature (below -60°C).
- ☆ The refrigerants used at present in cryogenic freezing are liquid nitrogen and liquid carbon dioxide.

Freezing may be achieved by

i. Immersion in the liquid
ii. Spraying of liquid
iii. Circulation of its vapour over the product to be frozen

(4) Dehydro-freezing

- ☆ Freezing is done after the partial dehydration.
- ☆ About 50 per cent of the moisture is removed by dehydration prior to freezing.
- ☆ This has been found to improve the quality of the food.
- ☆ Dehydration does not cause deterioration and dehydro-freezing foods are relatively more stable.

- ☆ Gamma rays or electron beams pass through foods there are collisions between the ionizing radiation and food particles at atomic and molecular levels, resulting in the production of ion pairs and free radicals.

(5) Freeze Drying (Sublimation-drying or lyophilisation)

In this process food is frozen first at -18°C on trays in the lower chamber of a freeze drier and the frozen material dried (initially at 30°C for 24hrs and then at 20°C) under high vacuum (0.1mm Hg) in the upper chamber. Direct sublimation of the ice takes place without passing through the intermediate liquid stage. The product is highly hygroscopic, excellent in taste and flavour and can be reconstituted readily. Mango pulp, orange juice concentrate, passion fruit juice and guava pulp are dehydrated by this method.

Preservation by Irradiation

Sterilization of food by ionizing radiations is a recently developed method of preservation which has not yet gained general acceptance. The unacceptable flavour of some irradiated foods and the fear that radioactivity might be induced in such food has come in the way of its greater use.

When gamma rays (or) electron beams pass through foods there are collisions between the ionizing radiation and food particles at atomic and molecular levels, resulting in the production of ion pairs and free radicals. The reactions of these products among themselves and with other molecules results in physical and chemical phenomena which inactivate microorganism in the food. Thus irradiation of food can be considered to be a method of 'Cold sterilization' *i.e.* food is free of microorganism without high temperature treatment. Radiation dose of upto 1 Mrad is not hazardous.

Ionizing radiations can be used for sterilization of foods in hermetically sealed packs, reduction of the spoilage organisms in the perishable foods, delays ripening of fruits, inhibits sprouting of root vegetables and controls infestation (insects) in stored cereals.

International unit of measurement - Gray (Gy) and symbol is radura

Based on Dose of Radiation

1. Low Dose - Radicidation

- ☆ Upto 1KGy
- ☆ Sprout inhibition in bulbs and tubers 0.03- 0.15KGy
- ☆ Delay in fruit ripening 0.25 – 0.75KGy
- ☆ Insect disinfestations and elimination of food borne parasites 0.07- 1.00KGy

2. Medium Dose - Radurization

- 1KGy to 10KGy
- Delay mold growth on strawberries at 1.00 – 1.50 KGy
- Reduction of spoilage microbes to prolong shelf life of meat, poultry and sea foods under refrigeration 1.50-3.00K
- Reduction of pathogenic microbes in fresh and frozen meat, poultry and sea food at 3.00-7.00KGy
- Reducing the number of microorganisms in spices 10.00KGy

3. High Dose - Radappertization

- Above 10KGy
- Sterilization of packaged meat, poultry and their products 25.00-75.00KGy
- Sterilization of hospital diets 25.00-75.00KGy
- Product implementation as increased juice yield or improved rehydration.

Chapter 12

Processing of Puffed Flaked and Extruded Products

Puffing

Puffing is a process of sudden release of water vapour and expansion of pregelatinised kernals. Cereals and millets such as rice, wheat, corn, sorghum, ragi are used for puffing.

Types of Puffing

- Sand roasting
- Oven puffing
- Gun puffing
- Extruded gun puffing
- HTST Fluidized bed puffing
- Microwave puffing

Gun-Puffed Whole Grain Cereals

Rice and wheat are the only types of grain used in gun-puffed whole grain production. In this process grain is loaded into the opening of gun and the lid is closed and sealed. As the gun begins to rotate, gas burners heat the sides of the gun body causing the moisture in the grain to convert to steam. When the lid is opened, the sudden change in pressure causes the grain to puff.

Whole grains

↓

Soaking loaded in gun chamber

↓

Heat by steam or gas burner

↓

Suddenly open the lid

↓

Puffing

↓

Drying

↓

Packing

Extruded Gun-Puffed Cereals

Extruded gun-puffed cereals use a meal or flour as the starting ingredient. The dough cooks in the extruders and is then formed into the desired shape through a die. After the dough is extruded, it is dried and tempered. It then undergoes the same puffing and final processing steps as described for whole grain gun-puffed cereals.

Flour

↓

Dough

↓

Extrude into desired shape

↓

Drying

↓

Gun puffing

Puffed Rice/Parched/Expanded Rice Processing

Paddy

↓

Soaking in water (30 per cent m.c)

↓

Parboiled rice (14 per cent m.c)

↓

Milling (15 per cent m.c)

↓

Dry heat parboiled rice in pans (110°C)

↓

Salting (Sodium chloride 10g/100 ml water per kg of rice)

↓

Roasing in hot sand (250°C)

↓

Expanded rice (0.5 per cent m.c)

Flaking

Flaking is a process of flattening the cooked grains by passing through revolving roller. Before flaking, grains and pulses are softened by being partially cooked in steam. They are then pressed or rolled into flakes which are subsequently dried. Such partially-cooked cereals may be used as quick-cooking or ready-to-eat foods.

Changes in Flaking

- ✰ Starch will get gelatinised and hydrolysed slightly
- ✰ Browning will take place due to the interaction between sugar and protein
- ✰ Enzyme process are stopped and holding the final product more stable
- ✰ Dextrination and caramelization is occur due to the high temperature in the roasting oven
- ✰ Flakes become crisper due to the very low moisture.

Flaked Rice Processing

Paddy

↓

Soaking (12-18 hours)

↓

Roasting (200-250°C for 20-30 sec)

↓

Tempering (5 min)

↓

Flaking (using an edge runner)

↓

Flaking machine

↓

Flaked rice

Extruded Products

Extrusion is a process that combines several unit operations, including mixing, kneading, shearing, heating, cooling, shaping and forming. It involves compressing and working raw materials *e.g.* flours, starches, proteins, salt, sugar and other minor ingredients, to form a semi-solid mass under a variety of controlled conditions and, then forcing into to pass through a restricted opening such as a shaped hole or slot at a predetermined rate. Heat is applied directly by steam injection or indirectly through a heated barrel or the conversion of mechanical energy. The final process temperature in the cooking extruder can be as high as 200oC but the residence time is relatively short *i.e.* 10-60 seconds. Thus extrusion cooking is also called a high temperature short time (HTST) process. This process of HTST, extrusion bring gelatinization of starch, denaturation of proteins, modification of lipids and inactivation of enzymes, microbes and many anti-nutritional factors.

Examples of Extruded Foods

Pasta products (noodles, vermicelli, spagheti)

Classification of Extruders

- Extruders are classified into two types according to operation: **Hot and cold extruders**
- Based on type of construction extruders are classified into: **Single screw and twin screw extruder**

Manufacturing of Noodles

Ingredients Required

Maida flour -100 gm,

Salt - 2 gm,

Water -35 ml

The sieved maida flour was steamed for 10 -15 minutes. The mixture was sieved again, fed into the extruder by selecting the die and conditioned by adding the required quantity of water and salt. The dough was extruded. The extruded products were steam cooked for 5 -10 minutes and tempered for 10 hrs. The tempered product was dried at 50°C for 2 - 4 hrs in a cabinet drier, cooled and packed in polyethylene pouches. It is extremely important to use the right kind of flour for the right product combination. Macroni, spaghetti and vermicelli belong to a class them "Pasta Alimentare" (Alimentary pasta, while German terminology is "Teigwaren" (pasta goods). The Macroni industry is well developed in Italy, USA, France, Switzerland and other Western Countries. Although vermicelli (Semia) has been known and produced in India for a long time, the production of macroni and spaghetti is of recent origin in the country.

Flow Chart

Maida flour (1kg) → Sieving (BS 60 sieve) → Steaming (10 -15 min.) → Sieving (BS 60 sieve) → Adding into the extruder → Wetting with water (350 - 400ml/kg) and salt (2g/kg) → Kneading (15 - 20min.) and extruding → Gelatinizing (steaming for 10 min.) Tempering overnight (10 hrs.) → Drying (50°C for 2 - 4 hrs) → Cooling and Packing.

Pasta Products Cooked with Vegetables

Ingredients

Noodles: 100g

Carrot: 25g

Beans: 25g

Cabbage: 25g

Big onion: 10g

Capsicum: 10g

Onion stalk: 5g

Coriander leaves: 5g

Soy sauce: 5ml

Tomato sauce: 3ml

Pepper: 5g

Salt: 5g

Oil: 10ml

Method

- The noodles were cooked with excess water, drained, cooled and mixed with soya and tomato sauce.
- To the oil in the frying pan onion was added and roasted well.
- To this vegetables and salt were added and cooked for 3 minutes on high flame.
- The cooked pasta and pepper was added to the pan, mixed well and further cooked for 2 minutes.
- The product was garnished with coriander leaves and served.

Cooking Quality of Noodles

The noodles (5g) was precooked for 30 minutes and cooked in 30ml of distilled water until the white core disappeared as determined by squeezing the noodles between glass slides and the cooking time was noted. The cooked noodles were placed in cold water, drained, wiped using paper towels and the gain in weight (water uptake) and cooked volume was recorded. Solid content in the cooked water was determined by boiling to dryness in water bath at 80°C followed by drying (6 hours) in a cabinet drier at 105°C.

Factors Affecting Extrusion Cooking

The two factors that most influence the nature of the extruded product are the rheological properties of the food and the operating conditions of the extruder.

Rheological Properties of the Food

The properties of the feed material have an important influence on the texture and colour of the product; the most important factors are:

1. The type of feed materials
2. Their moisture content
3. The physical state of the materials
4. Their chemical composition, particularly the amounts and types of starches, proteins, fats and sugars
5. The pH of the moistened material.

Operating Characteristics

The most important operating parameters in an extruder are:

1. Temperature
2. Pressure
3. Diameter of the die apertures
4. Shear rate.

Advantages of Extruded Products

1. Loss of nutrients during processing is minimized
2. There is flexibility to add desirable ingredient(s) in the product. Hence possible to produce nutritionally rich product.
3. It is easy to have desirable shape and size of the product
4. Broken rice, a by-product instead of whole rice can be used for making these products
5. It is a continuous process hence capacity is high

Chapter 13

Preservation by Using Sugar and Salt

I. Jam

Jam is a product made by boiling fruit pulp with sufficient quantity of sugar to a reasonably thick consistency, firm enough to hold the fruit tissues in position. Apply, sapota, papaya, plums, mango, grapes, jack, pineapple, banana, guava and pears are used for preparation of jam. It can be prepared from one kind of fruit or from two or more kinds. In its preparation about 45 per cent of fruit pulp should be used for every 55 per cent of sugar. The FPO specification of jam is 68.5 per cent TSS, 45 per cent of fruit pulp and 0.5-0.6 per cent of acid (citric acid) per 100 gm of the prepared product.

Preparation of Jam

a) Selection and Preparation of Fruit

Select good quality ripe fruits. Wash the fruits well in cold water. Peel the fruits and remove the stones and corers present. Cut the peeled fruit into small pieces with a stainless steel knife. If the fruit is hard, it should be cut into very small pieces. Pulp the fruits by using pulper.

b) Addition of Sugar and Acid

c) Cooking

Cook the mixture slowly with occasional stirring. The fruit pulp should be crushed with a laddle during cooking. Continue cooking till the temperature of the mass reaches 105.5°C.

Sheet (or) Flake Test

A small portion of jam is taken out during boiling in a spoon or wooden laddle and cooled slightly. It is then allowed to drop. If the product falls off in the form of a sheet (or) flakes instead of flowing in a continuous stream (or) syrup, it means that the end point has been reached and the product is ready. Otherwise boiling is continued till the sheet test is positive.

d) Packaging

Fill the hot jam into clean dry sterilized jars. Allow the jam to cool and fix the sterilized lid to the jar. Store it in a cool place.

Process

Ripe firm fruits → Washing → Peeling →Pulping (Remove seed and core) → Addition of sugar and acid → Boiling (with continuous stirring) → Judging of end point by further cooking up to 105°C (or) 68 per cent TSS (or) by sheet test → Filling hot into sterilized bottles → cooling → Sterilized bottles → cooling → Waxing → Capping → Storage (at ambient temperature).

Problems in Jam Production

i) **Crystallization:** The final product should contain 30 to 50 per cent invert sugar. If the percentage is less than 30, cane sugar may crystallize out on storage and if it is more than 50 the jam will become a honey like mass due to the formation of small crystals of glucose. Corn syrup or glucose may be added along with cane sugar to avoid crystallization.

ii) **Sticky (or) gummy jam:** Because of high percentage of TSS, jams tend to become gummy or sticky. This problem can be solved by addition of pectin (or) citric acid (or) both.

iii) **Premature setting:** This is due to low TSS and high pectin content in the jam and can be prevented by adding more sugar.

iv) **Surface graining and shrinkage:** This is caused by evaporation of moisture during storage of jam. Storing in a cool place can reduce it.

v) **Microbial spoilage:** Moulds may spoil the jam during storage but they are destroyed if exposed to less than 90 per cent humidity. Hence, jams should be stored at 80 per cent humidity. It is also advisable to add 40 ppm SO_2 in the form of KMS (or) 200 ppm of benzoic acid.

II. Jelly

A jelly is a semi solid product prepared by boiling a clear, strained solution of pectin containing fruit extract, free from pulp, after the addition of sugar and acid. A perfect jelly should be transparent, well set but not too stiff, and should have the original flavour of the fruit. It should be of attractive colour and keep its shape when removed from the mould. It should be firm enough to retain a sharp edge but tender enough when it is pressed. It should not be gummy, sticky or syrupy or have crystallized sugar. The product should be free from dullness with little (or) no syneresis (weeping) and neither tough nor rubbery. The FPO specification for jelly is the final product should have 65 per cent solids, 45 per cent fruit extract and 0.5-0.75 per cent acid.

Guava, sour apple, plum, wood apple, papaya and jack fruit are rich in pectin and generally used for preparation of jelly. Pineapple, strawberry grapes *etc.* can be used but only after addition of pectin powder, because these fruits have low pectin content. Preparation of jelly is similar to that of jam.

Process

Fruit (Firm, not over ripe) → Washing → Cutting into thin slices → Boiling with water (1 ½ times the weight of fruits for about 20-30 min) → Addition of citric acid during boiling (2 g per kg of fruit) → Straining of extract → Pectin test (for addition of sugar) → Addition of sugar → Boiling → Judging of end point (sheet/drop/temp test) → Removal of scum (or) foam (one teaspoonful of edible oil added for 45 kg sugar) → Addition of colour and remaining citric acid → Filling hot into clean sterilized bottles → Waxing (paraffin wax) → Capping → Storage at ambient temperature.

Important Considerations in Jelly Making

Pectin, acid, sugar (65 per cent) and water are the four essential ingredients. Pectin test and determination of end point of jelly formation are very important for the quality of jelly.

A. Pectin

Pectin substances present in the form of calcium pectate are responsible for the firmness of fruits. Pectin is the most important constituent of jelly. It is a commercial term for water soluble pectinic acid which under suitable conditions forms a gel with sugar and acid. In the unripe fruits, the pectic substance is present in the form of water insoluble protopectin. In the ripened fruits, with the help of protopectinase enzyme it is converted into pectin. In the over ripe fruits due to the presence of pectic methyl esterase (PME) enzyme, the pectin gets largely converted to pectic acid which is water insoluble. This is one of the reasons that both immature and over ripe fruits are not suitable for making jelly and only ripe fruits are used.

The setting of pectin is also dependent upon the pH and sugar concentrations. Stiffness of the gel increases with increasing concentration of pectin upto a certain point beyond which the addition of more pectin has little effect. Too little pectin gives soft syrup instead of gel. The jellying power of fruit pectin depends upon the amount of pectin used as well as its degree of polymerization and acetyl content.

The amount of pectin extracted varies with the method of extraction, the ripeness of the fruit, the quantity of water added for extracting the juice and the kind of fruit. Usually about 0.5-1.0 per cent of pectin of good quality in the extract is sufficient to produce good jelly. If the pectin content is higher a firm and tough jelly is formed and if it is less the jelly may fail to set.

Jelly Strength

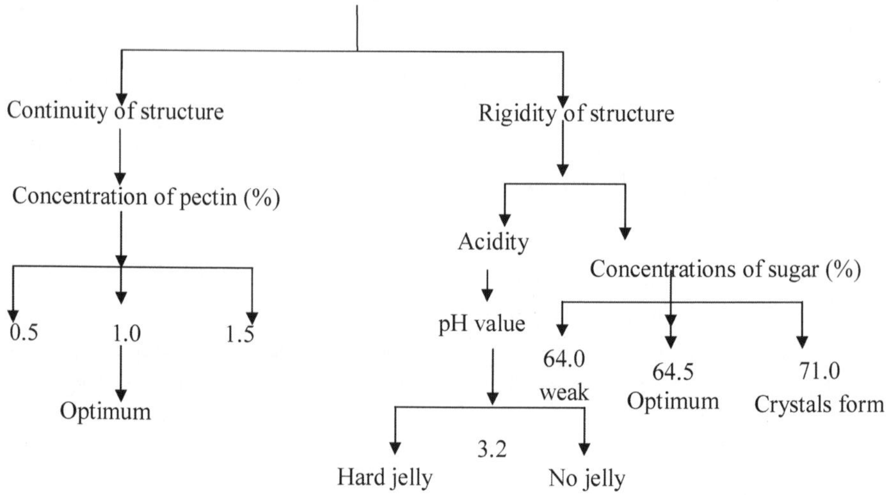

Determination of Pectin Content

The pectin content of the strained extract is usually determined by one of the following two methods

i) Alcohol test: This method involves precipitation of pectin with alcohol. One teaspoonful of strained extract is taken in a beaker and cooled, and 3 tea spoon full of methylated spirit are poured gently down the sides of the beaker which is retarded for mixing and allowed to stand for a few minutes.

a. If extract is rich in pectin, a single transparent lump (or) clot will form. An equal amount of sugar is to be added to the extract for preparation of jelly.

b. If extract contains a moderate amount of pectin, the clot will be less firm and fragmented. Three fourths the amount of sugar is to be added.

c. If extract is poor in pectin, numerous small granular clots will be seen. One half the amount of sugar is added.

ii) Jelmeter test: The jelmeter is held in the left hand with the thumb and forefinger. The bottom of the jelmeter tube is closed with the little finger. The strained extract is poured into the jelmeter with a spoon, held in the right hand, till it is filled to the brim. While still holding the jelmeter, the little finger is removed from the bottom end and the exactly is allowed to flow or drip for extract one minute, at the end of which the finger is replaced. The reading of the level of extract in the jelmeter is noted. This figure indicates how many parts of sugar are to be added to one part of juice.

B. Acid

The jellying of extract depends on the amount of acid and pectin present in the fruit. Of the three acids citric, malic and tartaric found in fruits, tartaric acid gives the best results. The final jelly should contain atleast 0.5 per cent but not more than 1 per cent total acids because a larger quantity of acid may cause syneresis.

pH of Extract

Jelly strength increases with the increase in pH until an optimum is reached. Further the addition of acid decreases the jelly strength. The optimum pH for a jelly containing 1per cent pectin is approximately 3.0, 3.2 and 3.4 for 60, 65 and 70 per cent TSS, respectively. The pH of the jelly can be controlled by 1) adjusting pH of extract with acid/alkali and (2) adding a suitable buffer. In general, the optimum pH value for jelly is 3.2.

C. Sugar

This essential constituent of jelly imparts to it sweetness as well as body. If the concentration of sugar is high, the jelly retains less water resulting in a stiff jelly, probably because of dehydration.

Judging of End Point

Boiling of jelly should not be prolonged, because excessive boiling results in a greater inversion of sugar and destruction of pectin. The important point to remember is that it is the fruit extract which requires boiling and not the added sugar. If a jelly is cooked for a prolonged period, it may become gummy, sticky, syrupy and deteriorate in colour and flavour. The end point of boiling can be judged in the following way.

1. **Sheet (or) Flake test:** As described under jam.
2. **Drop test:** A drop of the concentrated mass is poured into a glass containing water. Settling down of the drop without disintegration denotes the end point.
3. **Temperature test:** A solution containing 65 per cent TSS boils at 105°C. Heating of the jelly to this temperature would automatically bring the concentration of solids to 65 per cent. This is the easiest way to ascertain the end point.

Problems in Jelly Making

I. **Failure to set:** This may be due to

a. Addition of too much sugar
b. Lack of acid (or) pectin
c. Cooking below the end point
d. Cooking beyond the end point
e. Prolonged cooking.

II. **Cloudy (or) Foggy Jellies:** It is due to the following reasons

a. Use of non-clarified juice (or) extract

b. Use of immature fruits

c. Over-cooking

d. Over-cooling

e. Non-removal of scum

f. Faulty pouring

g. Premature gelation – due to excess of pectin in the extract

III. **Formation of crystals:** It is due to excess of sugar.

IV. **Syneresis (or) Weeping of jelly:** The phenomenon of spontaneous exudation of fluid from a gel is called syneresis (or) weeping. This may be due to

a. Excess of acid

b. Too low concentration of sugar

c. Insufficient pectin

d. Premature gelation

e. Fermentation.

III. Marmalade

This is a fruit jelly in which slices of the fruit (or) its peel are suspended. The term is generally used for products made from citrus fruits like oranges and lemons in which shredded peel is used as the suspended material. Citrus marmalades are classified into (1) jelly marmalade (2) jam marmalade. The FPO specifications for marmalade are TSS - 65 per cent and fruit juice - 45 per cent of the prepared product.

Ingredients

Pectin extract: 1 litre

Sugar: 750 gm

Shredded peel: 62 gm

Jelly marmalade: It prepared from the clarified pectin extract.

Process

Ripe fruits → Washing → Peeling outer yellow portion (Flavedo) thinly → Cutting yellow portion into fine shreds (1.9 - 2.5 cm long and 0.8 – 0.12 cm thick) → Cutting of 0.3-0.45 cm thick slices of peeled fruit (or) crushing into pulp in a greater → Boiling (in 2-3 times its weight of H_2O for 40-60 min.) Straining the extract → Testing for pectin content (alcohol test) → Addition of sugar (as required) → Cooking to 103-105°C (Continuous stirring) → Addition of prepared shreds (shredded peel boiled for 10 to 15 min. in several changes of water for softening and removing bitterness and added about 62 g per kg of extract) → Boiling till jellying point (continuous stirring) → Testing for end point (sheet/drop/temp. test) → Cooling (to 82-88°C with continuous stirring) → Flavouring (orange oil) → Filling in sterilized bottles → Sealing → Storage at ambient temperature.

Jam Marmalade

The method of preparation is practically the same as that for jelly marmalade. In this case the pectin extract of fruit is not clarified and the whole pulp is used. Sugar is added according to the weight of fruit, generally in the proportion of 1:1. The pulp - sugar mixture is cooked till the TSS content reaches 65 per cent.

Problems in Marmalade Making

Browning during storage is very common which can be prevented by addition of 0.09 gm of KMS per kg of marmalade and not using tin containers. KMS dissolved in a small quantity of water is added to the marmalade while it is cooling. KMS also eliminates the possibility of spoilage due to moulds.

2) Preservation with Salt

Salt improves the taste and flavour and hardness the tissues of vegetables and controls fermentation. Salt content of 15 per cent or above prevents microbial spoilage. This method of preservation is generally used only for vegetables which contain very little sugar and hence sufficient lactic acid cannot be formed by fermentation to act as preservative. However, some fruits like lime, mango, *etc.*, are also preserved with salt.

Pickling and Brining

The preservation of food in common salt or in vinegar is known as pickling. It is one of the most ancient methods of preserving fruits and vegetables. Pickles are good appetizers and add to the palatability of a meal. They stimulate the flow of gastric juice and thus help in digestion. Several kinds of pickles are sold in the Indian market. Mango pickle ranks first followed by cauliflower, onion, turnip and lime pickles. These are commonly made in homes as well as commercially manufactured and exported. Fruits are generally preserved in sweetened and spiced vinegar, while vegetables are pre- served in salt.

Pickling is the result of fermentation by lactic acid-forming bacteria" which are generally present in large numbers on the surface of fresh vegetables and fruits. These bacteria can grow in acid medium and in the presence of 8-10 per cent salt solution, whereas the growth of a majority of undesirable organisms is inhibited -. Lactic acid bacteria are most active at 30°C, so this temperature must be maintained as far as possible in the early stage of pickle making. When vegetables are placed in brine, it penetrates into the tissues of the former and soluble material present in them diffuses into the brine by osmosis, The soluble material includes fermentable sugars and minerals. The sugars serve as food for lactic acid bacteria which convert them into lactic and other acids. The acid brine thus formed acts upon vegetable' tissues to produce the characteristic taste and aroma of pickle. In the dry salting method several alternate layers of vegetables and salt (20- 30 g of dry salt per kg vegetables) are kept in a vessel which is covered with a cloth and a wooden board and allowed to stand for about 24 hours. During this period, due to osmosis, sufficient juice comes out from the vegetables to form brine. Vegetables which do not contain enough juice (*e.g.*, cucumber) to dissolve the added salt are covered with brine (steeping in a concentrated salt solution is known as brining). The amount of brine required is usually equal to half the volume of vegetables. Brining is the most important step in pickling. The growth of a majority of spoilage organisms is inhibited by brine containing 15 per cent salt. Lactic acid bacteria, which are salt-tolerant, can thrive in brine of 8-10 per cent strength though fermentation takes place fairly well even in 5 per cent brine. In a brine, containing 10 per cent salt fermentation proceeds somewhat slowly. Fermentation takes place to some extent up to 15 per cent but stops at 20 per cent strength. It is, therefore, advisable to place the vegetables in 10 per cent salt solution for vigorous lactic acid fermentation.

Methods of Pickling

Pickling uses the salt combined with the acid, such as acetic acid (vinegar), oil and spices. Based on the ingredients, the pickles differ. Different methods of pickling are:

- Pickles with salt
- Pickles with vinegar
- Pickles with oil
- Pickles with salt, vinegar, oil and spices

Pickles with Salt

Vegetables are usually preserved by salting. Salt content of 15 per cent or above prevents microbial spoilage as well as inhibits enzymatic browning and discolouration. Salting improves the taste and renders firmness in tissues of vegetables and controls fermentation. This method is suitable for vegetables as they contain very little sugar, as a result lactic acid is not formed in sufficient amounts to act as preservative. However, some fruits *viz.* mango, lemon, *etc.* are also preserved with salt.

Pickles with vinegar

Vinegar pickles are the most important pickles consumed in other countries. Vinegar in concentrations up to 4 per cent acetic acid and higher concentrations has bacteriostatic and bactericidal action, respectively. The final concentration of 2-3 per cent acetic acid is required to ensure preservation. For this, 6-9 per cent acetic acid vinegar is used, to overcome dilution by water oozing out of tissues. Such higher concentration helps to expel the gases present in the intercellular spaces of vegetable tissue. In vinegar pickles, salt (2-3 per cent) and sometimes sugar (2-5 per cent) are also added. If the vinegar concentration is lower than 2 per cent, vinegar pickles need to be subjected to pasteurization in order to assure their preservation. Mango, garlic, chilies, *etc.* are preserved in vinegar.

Pickles with Oil

Highly spiced oil pickles are quite popular in India. Oils like mustard oil, rapeseed oil, sesame oil are generally used for pickling. The fruits or vegetables should be completely immersed in the edible oil. Cauliflower, lime, mango and turnip pickles are the most important oil pickles. The pickle remains in good condition for one to two years if handled properly.

Pickles with Salt, Vinegar, Oil and Spices

This method combines the advantages of fermentative action of salt and the preservation action of both vinegar and oil. The flavouring and antibacterial property of spices is also made use of in this kind of pickle. The powdered spices are usually fried in oil and mixed to the prepared fruit/vegetable before the addition of vinegar.

Pickling Procedure

Cucumbers

↓

Washing

↓

Peeling

↓

Cutting into 5 cm pieces

↓

Mixing with salt

↓

Filling in jar

↓

Standing for 6-8 hours

↓

Draining off water

↓

Adding spices and vinegar

↓

Keeping in sun for a week

↓

Storage

Mangoes
(Mature, green)
↓
Washing
↓
Peeling
↓
Slicing
↓
Putting slices in jar
Sprinking salt
↓
Putting in sun for one week
(Shaking jar at least twice a day to mix the salt)
↓
Mixing spices
↓
Storage at ambient temperature (in cool and dry place)

Problems in Pickle Making

1. **Bitter taste**: Use of strong vinegar or excess spice or prolonged cooking of spices imparts a bitter taste to the pickle
2. **Dull and faded product**: This is due to use of inferior quality materials or insufficient curing.
3. **Shrivelling**: It occurs when vegetables (*e.g.* cucumber) are placed directly in a very strong solution of salt or sugar or vinegar. Hence, a dilute solution should be used initially and its strength gradually increased.
4. **Scum formation**: When vegetables are cured in brine, a white scum always form on the surface due to the growth of wild yeast. This delays the formation of lactic acid and also helps the growth of putrefactive bacteria which cause softness and slipperiness. Hence, it is advisable to remove scum as soon as it is formed. Addition of one per cent acetic acid helps to prevent the growth of wild yeast in brine, without affecting lactic acid formation.
5. **Softness and slipperiness**: This very common problem is due to inadequate covering with brine or the use of weak brine: The problem

can be solved by using a brine of proper strength and keeping the pickles well below the surface of the brine

6. **Cloudiness**: When the structure of the vegetable used in pickling, *e.g.*, onion, is such that the acetic acid (vinegar) cannot penetrate deep enough into its tissues to inhibit the activity of bacteria and other microorganisms present in them, fermentation starts from inside the tissues; rendering the vinegar cloudy. This microbial activity can only be checked by proper brining. Cloudiness may also be caused by use of inferior quality vinegar or chemical reaction between vinegar and minerals.
7. **Blackening**: It is due to the iron in the brine or in the process equipment reacting with the ingredients used in pickling. Certain microorganisms also cause blackening.

Chapter 14

Canning of Fruits and Vegetables

The process of sealing food stuffs hermetically in containers and sterilizing them by heat for long storage is known as canning. In 1804, Appert in France invented a process of sealing foods hermetically in containers and sterilizing them by heat. In honour of the inventor, canning is also known as appertizing. Saddington in England was the first to describe a method of canning of foods in 1807.

In 1810, Peter Durand, another Englishman, obtained the first British Patent on canning of foods in tin containers. In 1817, William Underwood introduced canning of fruits on a commercial scale in U.S.A. Fruits and vegetables are canned in the season when the raw material is available in plenty. The canned products are sold in the off-season and give better returns to the grower.

Principles and Process of Canning

Principle of Canning

Destruction of spoilage causing micro-organisms and enzymes within the sealed containers by means of heat.

Machinery/Equipments Required for Canning

- Can Reformer, Flanger, Double Seamer, Exhaust box, Sterilization tank, Retort, Lye
- Peeling tank and Coding Machine, Refractrometer, gas Boiler.
- Empty flat cans of A1 tall or A21/2 size.
- Fruits and vegetables suitable for canning
- Peeling and coring knives

- ☆ Salt or sugar
- ☆ Sodium hydroxide

Process

- ☆ Selection of fruits/vegetables
- ☆ Grading
- ☆ Washing
- ☆ Cooling
- ☆ Blanching
- ☆ Cutting
- ☆ Peeling
- ☆ Filling and Syruping or Brining
- ☆ Exhausting
- ☆ Storage
- ☆ Cooling
- ☆ Processing
- ☆ Sealing

1. Selection of Fruits and Vegetables

- ☆ Fruits and vegetables should be absolutely fresh.
- ☆ Fruits should be ripe, but firm, and uniformly mature.
- ☆ Over-ripe fruits should be rejected because they are infected with microorganisms and give a poor quality product.
- ☆ Unripe fruits should be rejected because they generally shrivel and toughen on canning.
- ☆ All vegetables except tomatoes should be tender.
- ☆ Tomatoes should be firm, fully ripe and of deep red colour.
- ☆ Fruits and vegetables should be free from dirt.
- ☆ They should be free from blemishes, insect damage or mechanical injury.

2. Grading

The selected fruits and vegetables are graded according to size and colour to obtain uniform quality. This is done by hand or by machines such as screw grader and roller grader.

Fruits like berries, plums and cherries are graded whole, while peaches, pears, apricots, mangoes, pineapple, *etc.*, are generally graded after cutting into pieces or slices.

3. Washing

It is important to remove pesticide spray residue and dust from fruits and vegetables. One gram of soil contains 1012 spores of microorganisms. Therefore, removal of microorganisms by washing with water is essential. Fruits and vegetables can be washed in different ways. Root crops that loosen in soil are washed by soaking in water containing 25 to 50 ppm chlorine (as detergent). Other methods of washing are spray washing, steam washing, *etc.*

4. Peeling

The objective of peeling is to remove the outer layer. Peeling may be done in various ways.

Hand Peeling

It is done mostly in case of fruits of irregular shape,

e.g. mango and papaya, where mechanical peeling is not possible.

Steam Peeling

Free-stone and clingstone peaches are steam peeled in different ways. The former are cut and steam washed. Potatoes and tomatoes are peeled by steam or boiling water.

Mechanical Peeling

This is done in case of apples, peaches, pineapples and cherries and also for root vegetables like carrots, turnips and potatoes.

Lye Peeling

Fruits like peaches, apricots, sweet oranges, mandarin oranges and vegetables like carrots and sweet potatoes are peeled by dipping them in 1 to 2 per cent boiling caustic soda solution (lye) for 30 seconds to 2 minutes depending on their nature and maturity. Hot lye loosens the skin from the flesh by dissolving the pectin. The peel is then removed easily by hand. Any trace of alkali is removed by washing the fruit or vegetable thoroughly in running cold water or dipping it for a few seconds in 0.5 per cent citric acid solution. This is a quick method where by cost and wastage in peeling is reduced.

Flame Peeling

It is used only for garlic and onion which have a papery outer covering. This is just burnt off. Vegetables like peas are shelled, carrots are scarped, and beans are snipped or trimmed.

5. Cutting

Pieces of the size required for canning are cut. Seed, stone and core are removed. Some fruits like plum from which the seeds cannot be taken out easily are canned whole.

6. Blanching

It is also known as scalding, parboiling or precooking. It is usually done in case of vegetables by exposing them to boiling water or steam for 2 to 5 minutes, followed by cooling. The extent of blanching varies with the food. Generally fruits are not blanched.

This brief heat treatment accomplishes the following:

- Inactivates most of the plant enzymes which cause toughness, discolouration (polyphenol oxidase), mustiness, off-flavour (peroxidase), softening and loss of nutritive value.
- Reduces the area of leafy vegetables such as spinach by shrinkage or wilting, making their packing easier.
- Removes tissue gases which reduce sulphides.
- Reduces the number of microorganisms by as much as 99 per cent.
- Enhances the green colour of vegetables such as peas, broccoli and spinach.
- Removes saponin in peas.
- Removes undesirable acids and astringent taste of the peel, and thus improves flavour.
- Removes the skin of vegetable such as beetroot and tomatoes which helps in their peeling.

Disadvantages

- Water-soluble materials like sugar and anthocyanin pigments are leached by boiling water.
- Fruits lose their colour, flavour and sugar.

7. Cooling

After blanching, the vegetables are dipped in cold water for better handling and keeping them in good condition.

8. Filling

Before filling, cans are washed with hot water and sterilized but in developing countries these are subjected to a jet of steam to remove dust and foreign material. Automatic, large can- filling machines are used in advanced countries but choice grades of fruits are normally filled by hand to prevent bruising. In India, hand filling is the common practice. After filling, covering with syrup or brine is done and this process is called syruping or brining.

A 1-b butter size can should hold 230 285 g of fruit slices and a A 2 1D 2 size can 510 to 565 g.

The blanched vegetables are packed in sterilized cans which should hold the drained weight of vegetables as specified below:

1 lb butter size can - 269-283 g

A 2 1D 2 size can - 538-566 g

Pint size glass jar - 283-311 g

(i) Syruping

A solution of sugar in water is called syrup. Normally sucrose syrup is used in canning. Syrup is added to improve the flavour and to serve as a heat transfer medium for facilitating processing. Syruping is done only for fruits. Strained, hot syrup of concentration 20 to 55° Brix is poured on the fruit. Fruits rich in acid require more concentrated syrup than less acid ones. The syrup should be filled at about79 to 82°C, leaving a head space of 0.3 to 0.5 cm. Sometimes citric acid and ascorbic acid are also mixed with the syrup to improve flavour and nutritional value, respectively.

(ii) Brining

A solution of salt in water is called brine. The objective of brining is similar to that of syruping. Only vegetables are brined. Common salt of good quality free from iron should be used. Hot brine of 1 to 3 per cent concentration is used for covering vegetables and is filled at 79 to 82°C, leaving a head space of 0.3 to 0.5 cm. The brine should be filtered through a thick cloth before filling.

After syruping or brining the cans are loosely covered with lids and exhausted. Lidding has certain disadvantages such as spilling of the contents and toppling of the lids. Hence lidding has now been modernized by 'clinching' process in which the lid is partially seamed. The lid remains sufficiently loose to permit the escape of dissolved as well as free air from the can and also the vapour formed during the exhausting process.

9. Exhausting

The process of removal of air from cans is known as exhausting. After filling and lidding or clinching, exhausting is essential.

The major advantages of exhausting are as under:

a) Corrosion of the tinplate and pinholing during storage is avoided.

b) Minimizes discolouration by preventing oxidation.

c) Helps in better retention of vitamins particularly vitamin C.

d) Prevents bulging of cans when stored in a hot climate or at high altitude.

e) Reduces chemical reaction between the container and the contents.

f) Prevents development of excessive pressure and strain during sterilization.

Containers are exhausted either by heating or mechanically. The heat treatment method is generally used. The cans are passed through a tank of hot water at 82 to 87°C or move on a belt through a covered steam box. In the water exhaust box, the cans are placed in such a manner that the level of water is 4-5 cm below their tops. The exhaust box is heated till the temperature of water reaches 82 to 100°C and the centre of the can shows a temperature of about 79°C. The time of exhausting varies from 6 to 10 minutes, depending on the nature of the product. In the case of glass jars or bottles, vacuum closing machines are generally used. The bottles or jars are placed in a closed chamber in which a high vacuum is maintained.

It is preferable to exhaust the cans at a lower temperature for a longer period to ensure uniform heating of the contents without softening them into pulp. Exhausting at high temperature should be avoided because the higher the temperature, the more is the volume of water vapour formed, and consequently the greater the vacuum produced in the can.

10. Sealing

Immediately after exhausting the cans are sealed airtight by means of a can sealer. In case of glass jars a rubber ring should be placed between the mouth of the jar and the lid, so that it can be sealed airtight. During sealing the temperature should not fall below 74°C.

11. Processing

Heating of foods for preserving is known as processing, however, in canning technology processing means heating or cooling of canned foods to inactivate bacteria. Many bacterial spores can be killed by either high or every low temperature. Such drastic treatment, however affects the quality of food. Processing time and temperature should be adequate to eliminate all bacterial growth. Moreover, over-cooking should be avoided as it spoils the flavour as well as the appearance of the product.

Almost all fruits and acid vegetables can be processed satisfactorily at a temperature of 100°C, *i.e.*, in boiling water. The presence of acid retards the growth of bacteria and their spores. Further, they do not thrive in heavy sugar syrup which is normally used for canning of fruits. Vegetables (except the more acid ones like tomato and rhubarb) which are non-acid in nature, have a hard texture, and proximity to soil which many infect them with spore-bearing organisms processed at higher temperatures of 115 to 121°C.

The sourness of fruits and vegetables is due to their acid content (measured in pH) which has a great influence upon the destruction of microorganisms. The lower the pH the greater is the ease with which a product can be processed or sterilized.

Bacterial spores can be more easily destroyed at pH 3.0 (fruits) than at pH 5.0 to 6.0 (vegetables, except tomato and rhubarb). Bacterial spores do not grow or germinate below pH 4.5. Thus, a canned product having pH less than 4.5 can be processed in boiling water but a product with pH above 4.5 requires processing at 115 to 121°C under a pressure of 0.70 to 1.05 kg/cm^2 (10 to 15 lb/sq inch). It is essential that the centre of the can should attain these high temperatures.

The temperature and time of processing vary with the size of the can and the nature of the food: the larger the can, the greater is the processing time. Fruits and acid vegetables are generally processed in open type cooker, continuous non-agitating cookers, while vegetables (non-acid) are processed under steam pressure in closed retorts known as automatic pressure cookers and continuous agitating cookers. In India, small vertical stationary retorts (frontispiece) are generally used for canned vegetable processing. The sealed cans are placed in the cookers, keeping the level of water 2.5 to 5.0 cm above the top of the cans. The cover of the cooker is then screwed down tightly and the cooker heated to the desired temperature. The period of sterilization (process) should be counted from the time the water starts boiling. After heating for the required period the cooker is removed from the fire and the petcock is opened. When the pressure comes down to zero the cover is removed and the cans are taken out.

Fruits and vegetables can be classified into the following four groups according to their pH value

Class	*pH*	*Product*
Low acid (non-acid)	Above 5.0	Vegetables such as peas, lima bean, asparagus, cauliflower, potato, spinach, beet, corn, french bean
Medium acid	4.5-5.0	Turnip, carrot, okra, cabbage, pumpkin, beet, green bean, *etc.*, and products like soups and sauces
Acid	3.7-4.5	Tomato, pear, banana, mango, jackfruit, pineapple, sweet cherry, peach, apple and other fruits
High acid	Below 3.7	Citrus juice, rhubarb, prune, sauerkraut, pickle, chutney, *etc.*

12. Cooling

After processing, the cans are cooled rapidly to about 39°C to stop the cooking process and to prevent stack-burning. Cooling is done by the following methods.

(i) Dipping or immersing the hot cans in tanks containing cold water.

(ii) Letting cold water into the pressure cooker especially in case of vegetables.

(iii) Spraying cans with jets of cold water; and

(iv) Exposing the cans to air.

Generally the first method, *i.e.* dipping the cans in cold water, is used. If canned products are not cooled immediately after processing, peaches and pears becomes dark in colour, tomatoes turn brownish and bitter in taste, peas become pulpy with cooked taste and many vegetables develop flat sour (become sour).

13. Storage

After labelling the cans, they should be packed in strong wooden cases or corrugated cardboard cartons and stored in a cool and dry place. The outer surface of the cans should be dry as even small traces of moisture sometimes induce rusting. Storage of cans at high temperature should be avoided, as it shortens the shelf-life of the product and often leads to the formation of hydrogen swell.

The marketable life of canned products varies according to the type of raw materials used. Canned peach, grapefruit, pineapple, beans, spinach, pea, celery, *etc.* can be stored for about two years, while pear, apricot, carrot, beetroot, tomato, *etc.* can be stored for a comparatively long period.

FPO Specification for Canned Fruits and Vegetables

Specifications	*Canned Fruits*	*Canned Vegetables*
Drained weight	Not less than 50 per cent Exception: berry fruits 40 per cent	Not less than 55 per cent Exception: canned tomato 50 per cent
Appearance	Free from blemishes, stalks, leaves, *etc.*	Free from pods, stalks, roots and blemishes *etc.*
Texture	Free from disintegration and damage from bruises.	Free from disintegration and damage from bruises.
Added colour	No colour except strawberry and cherry	Not permitted, except for peas
Organoleptic quality	Characteristic taste	Characteristic taste
Preservatives	Not permitted	Not permitted
Incubation condition	No positive pressure when incubated at 37°C for one week	No positive pressure and sign of bacterial growth when incubated at 37°C and 55°C for one week.

Spoilage of Canned Products

Canned foods are liable to spoilage in storage for various reasons. The two important kinds of spoilages are:

1. Spoilage due to physical and chemical changes
2. Spoilage by micro-organisms

Various indications (Symptoms) of spoilage in cans are:

1. Swell

Swelled can, the ends are tightly bulged due to the formation of CO_2 or other gas inside the can as a result of decomposition of the contents by micro-organisms. The decomposed food in the can has an offensive and sour odour and is discolored. It is not fit for consumption and even may contain toxins produced by bacilli like *Clostridium botulinum*.

2. Hydrogen Swell

Hydrogen formed by the action of the acids present in the fruit on the tin plate causes the can to bulge at both the ends. In such cases, the food remains free from harmful micro-organisms and is still fit for consumption.

3. Springer

A mild swell at one or both ends of a can is called a 'Springer'. It may be an initial stage of hydrogen swell or may be caused by insufficient exhausting or by over filling of the can. Food generally remains fit for consumption.

4. Flipper

The cans appear normal, but when struck against a table top one or both ends become convex and spring or flip out, but can be pushed back to normal condition by a little pressure. Such a can is termed as 'flipper'. It may be an initial stage of swell or hydrogen swell, but does more frequently, owe to over filling or under exhausting.

5. Flat Sour

it is caused mostly in non-acid foods like vegetables by micro-organisms like thermophyllic bacteria (*B. coagulens* and *B. sterothermophilus*) which produce acid without formation of gas. The thermophyllic bacteria thrive at a high temperature of 100°C. If can are stored without adequate cooling, the contents remain at a temperature favourable for incubation of such bacteria for fairly long time, which results in their multiplication and spoil the product. It is therefore difficult to detect from the external appearance. It may be due to by under sterilization (processing). Therefore the thermophyllic bacteria would be of significance. The product has a sour odour, and its acidity much higher than that of the normal product. It is not fit for consumption.

6. Leakage

A leaking can is known as leaker. A very small leak may appear in the can owing to:

1. Defective seaming,
2. Nail holes caused by faulty nailing of cases while packing
3. Excessive internal pressure due to microbial spoilage sufficient to burst the can
4. Internal or external corrosion
5. Mechanical damage during handling.

7. Breathing (Breather)

There may be a very tiny leak in the can through which air may pass in and destroy the vacuum but not the micro-organisms. Consequently the vacuum in the can is always nil, and the pressure inside the can is equal to that of atmosphere. The damage to food is usually owing to rusting of the can caused by oxygen in the air but still the food remains fit for consumption.

8. Bursting of the Can

Cans may sometimes burst. This may be due to excess of pressure caused by the gas inside, produced by the decomposition of the food by micro-organisms, or by hydrogen gas formed by chemical action of acids of the food on the tin plate. Thus the canned product becomes a total loss.

Enzymatic Spoilage

Many reactions in plant and animal tissues are activated by enzymes. The changes in foods during storage can be produced both by enzymes present in the food or by enzymes from microorganisms that contaminate the food. A good example of the former is the ripening of banana due to the enzymes present which hasten the ripening process. After some time the fruit become too soft and unfit to eat. If there is a bruised spot on the fruit, yeasts can grow and produce enzymes which spoil the fruit.

Enzymes convert starch into sugars, protein into amino acids, and pectin into pectic acids and thus change the constituents of food. Some fruits and vegetables turn brown when damaged or when their cut surfaces are exposed to air due to the presence of the enzymes phenolase, peroxidase and polyphenol oxidase. Their actions can be easily controlled by regulating the temperature and excluding moisture and air. Enzyme can act between zero and 60°C. The optimum temperature of reaction is usually 37°C, the rate varying directly with temperature. All enzymes are inactivated at 80°C.

Microbial Spoilage

Bacteria, yeasts and moulds may infect food after harvesting, during its handling, processing and storage. But not all microorganisms cause spoilage, *e.g.*, lactic acid bacteria are used in the making of cheese and other fermented dairy products, yeasts for the production of wine and beer and *Acetobacter* bacteria for vinegar production. Spoilage organisms are present everywhere- in soil, air, water and even in the raw and processed food.

(i) Bacteria

These are unicellular microorganisms that are classed as plants though they do not contain chlorophyll. A bacterial cell is about 0.5 to 5 micrometer in length

Bacteria are classified according to their shape. *Cocci* are spherical, bacilli are cylindrical and *spirilla* and *vibrios* are spiral. Bacterial spores are more resistant than yeast or mould spores to most processing conditions. Bacteria, with a few exceptions, cannot grow in acid media in which yeasts and moulds thrive. They multiply by 'fission' or division of cells. When a bacterium becomes mature it divides into two, these two become four and so on. The growth of bacteria is very rapid and depends upon the nature of the food material, moisture, temperature and air. Some bacteria do not grow in air but temperature plays a major role in their growth, the optimum being generally 37°C.

Some bacteria produce spores which can be destroyed by heating at 121°C for 30-40 minutes. Bacteria are very sensitive to acids and are destroyed in their presence even at the temperature of boiling water. Hence, most fruits being acidic can be easily sterilized at 100°C whereas vegetables being non-acidic require a higher temperature of 116°C.

The important groups of bacteria are:

(a) *Bacillus*: rod-shaped
(b) *Coccus*: spherical
(c) *Coccobacillus*: oval-shaped
(d) Aerobes: require atmospheric oxygen for growth, *e.g., Acetobacter aceti*
(e) Facultative anaerobes: can grow with or without atmospheric oxygen
(f) Obligate anaerobes: do not grow in atmospheric oxygen
(g) Mesophiles: require a temperature below 38°C for growth
(h) Obligate thermophiles: grow between 38 and 82°C
(i) Facultative thermophiles: grow over the whole range of temperature covered by mesophiles and obligate thermophiles and below
(j Psychrotrophs: grow fairly well at refrigeration temperatures and some can even grow slowly at temperature below freezing.

Important Food Spoilage Bacteria

Group	***Genus***
Acetics	*Acetobacter and Gluconobacter*
Lactics	*Lactobacillus, Leuconostoc, Pediococcus, Streptococcus*
Butyrics	*Clostridium*
Propionics	*Propionobacterium*
Proteol Gytics	*Bacillus, Pseudomonas, Clostridium, Proteus etc.*

Some Useful Bacteria

The following bacteria are of great importance in the food processing industry.

Acetobacter sp.

These bacteria, also known as "Vinegar bacteria", cause significant spoilage in the wine industry but are necessary for vinegar production. The important species are *Acetobacter aceti, A. orleansis* and *A. Schutzenbachi*. They are very small, usually non-motile and generally do not form spores. These bacteria are aerobes and in the presence of oxygen convert ethyl alcohol to acetic acid. They are of two types- one type forms a tough shiny film on the surface of wine and the growth is known as "vinegar mother", while the other grows throughout the wine without forming "vinegar mother". These bacteria can be easily destroyed by heating to 65°C.

Lactobacillus sp.

Different organisms of this group, also known as "lactic acid bacteria", have different properties but all of them produce lactic acid from carbohydrates. Those which are used in distilling and brewing industries are facultative thermophiles (heat-tolerants) which grow abundantly at 50 to 55°C and produce much lactic acid. Mesophiles are used in the preparation of pickles. *Lactobacillus plantarum* is generally found in pickles and olives. The other important species are *Pediococcus cerevisiae, Leuconostoc mesenteroides, Streptococcus faecalis* and *Lactobacillus brevis*. These bacteria cause "lactic souring" and spoil wines, which can be easily prevented by maintaining a sulphur dioxide concentration of 0.007 per cent in wine.

(ii) Yeasts

Yeasts are unicellular fungi which are widely distributed in nature. They are somewhat larger than bacteria. The cell length is about 10 micrometer. Most yeasts are spherical or ellipsoidal. Yeasts that multiply by means of 'budding' are known as 'true yeasts'. The bud when it becomes mature separates from the mother cell and functions like an independent organism. Yeasts grow luxuriantly at a moderate temperature in a solution of sugar in plenty of water. Under suitable conditions the sugar is converted into alcohol and carbon dioxide gas is evolved.

Yeast + Sugar → Alcohol + Carbon dioxide

This is the reason that carbon dioxide is evolved from food materials spoiled by yeasts and pushes out corks from bottles with great force. Active fermentation can be easily recognized by the formation of carbon dioxide foams or bubbles. Yeasts prefer a low concentration of sugar for their growth. Most of them do not develop in media containing more than 66 per cent sugar or 0.5 per cent acetic acid. Boiling destroys the yeast cells and spores completely. Some of the yeasts which grow on fruits are *Saccharomyces, Candida* and *Brettanomyces.*

Pseudo-yeasts

These are like true yeasts but do not form spores. All the members of this group are particularly unsuitable for fermentation purposes as they produce off-flavours and cloudiness.

Yeasts Causing Food Spoilage

Yeast	*Product Spoiled*
Saccharomyces	Low-sugar products
Candida	High-acid foods, salty foods, butter
Brettanomyces	Beers, wines
Zygosaccharomyces (osmophilic)	Honey, syrups, molasses, wines, soy sauce
Pichia	Wines
Hansenula	Beers
Debaryomyces	Meat brine, cheese, sausages, *etc.*
Hanseniospora	Fruit juices
Torulopsis	Milk products, fruit juices, acid foods
Rhodotorula	Meat, sauerkraut
Trichosporon	Chilled beer

(iii) Moulds

Moulds are multicellular, filamentous fungi belonging to the division Thallophyta but are devoid of chlorophyll. They are larger than yeasts. They are strict aerobes and require oxygen for growth and multiplication and tend to grow more slowly than bacteria.

The principal parts of a mould are a web-like structure known as mycelium and the spore. The mycelium is often white and cottony and penetrates into the attacked foodstuff. After fixing itself the mould produces viable spores which resist the unfavourable conditions after dispersal and germinate when they get favourable conditions. They thrive best in closed, damp and dark situations with an adequate supply of warm, moist air but require less free moisture than yeasts and bacteria. They prefer sugar-containing substances and may spoil jams, jellies, preserves and other sugar-based products. Acid medium favours their growth and, therefore, they grow well in pickles, juices, *etc.* This is the main reason that fruits and fruit products are attacked by moulds which not only consume nutrients present in the food thereby lowering its food value but also spoil the flavour, texture and appearance of the product. They may grow even on moist leather but do not thrive in an alkaline medium. Moulds are sensitive to heat; boiling quickly destroys both moulds and their spores.

The most important moulds are:

a) *Penicillium* sp. (Blue moulds),

b) *Aspergillus* sp. (Black moulds),

c) *Mucor* sp. (Gray moulds), and

d) *Byssochlamys fulva*

a. *Penicillium sp.*

These are also known as blue moulds. In the initial stage of growth they have a cottony appearance but later when the spores or conidia are formed, their appearance becomes powdery and the colour becomes blue, brown or pink according to age. The spoiled materials have a 'mouldy' odour and flavour.

b. *Aspergillus sp.*

In the initial stages of growth it is white and cottony like Penicillium but later with the formation of spores, it becomes black and is hence known as "black mould". Unlike Penicillium it does not produce off-odour and flavour. They generally attacks grapes and bael.

c. *Mucor sp.*

It is gray in colour and hence is known as 'gray mould'. It is also known as 'pin mould' or bread mould' because if frequently grows on moist bread. Although Mucor attacks fruits, in the preservation of fruits and vegetables it does not pose a serious problem like blue and black moulds.

d. *Byssochlamys fulva*

This mould causes spoilage of canned fruits. The infected fruits disintegrate and sometimes carbon dioxide gas is also produced. It can grow under reduced oxygen tension and the ascospores posses' high resistance to heat. For destroying spores heating of the cans at 88 to 90°C is essential.

Although the organism is not as resistant as some of the thermopiles, its control in canned foods is difficult. Canned products which cannot withstand prolonged heating without deterioration are ultimately spoiled. Association of this mould with fruit in the field has been observed. Hence the emphasis should be on eliminating the organism from the raw material itself instead of processing to destroy it in the can.

A small number of moulds produce toxic substances, known as mycotoxins, in food. *Aspergillus flavus* produces aflatoxins in harvested crops, such as groundnut, which are stored in the field without drying properly.

Chapter 15

Drying and Dehydration

The practice of drying of food stuffs, specially fruits and vegetables, for preserving them is very old. The term 'drying' and 'dehydration' means the removal of water. The former term is generally used for drying under the influence of non-conventional energy sources like sun and wind. If fruits (or) vegetables are to be sun dried, they (or) their pieces should be evenly spread in single layer (on) trays or boards and exposed to the sun. In sun drying there is no possibility of temperature and humidity control. The hottest days in summer are, therefore, chosen so that the foods dry very fast, thus preventing them from getting spoiled due to souring. Souring (or) turning acidic is usually due to growth of microorganisms which convert the carbohydrates in the food to acid. Quick removal of moisture prevents the growth of the microorganisms.

Dehydration means the process of removal of moisture by the application of artificial heat under controlled conditions of temperature, humidity and air flow. In this process a single layer of fruits (or) vegetables, whole or cut into pieces (or) slices are spread on trays which are placed inside the dehydrator. The initial temperature of the dehydrator is usually 43°C which is gradually increased to 60-66°C in the case of vegetables and 50-71°C for fruits.

Advantages of Dried/Dehydrated Foods

- ☆ Dried foods are in more concentrated form than foods preserved in other ways. They are less costly to produce than canned or preserved food, because of lower labour costs and because of no sugar is required.
- ☆ Due to reduction in bulk of the product, it requires less storage space.
- ☆ The weight of a product is reduced to 1/4th to 1/9th its original (or) fresh weight and thus the cost of its transport is reduced.

Sun Drying

Sun drying of fruits and vegetables is practiced widely in tropical and subtropical regions where there is plenty of sun shine and practically little or no rain during the drying season. Sun drying in direct (or) diffused sunlight (shade drying), one of the earliest method of food preservation, is still used for the production of dried fruits, and also for drying nuts. It was originally limited to fruits high in sugar content, which when harvested, would dry naturally without hazard of loss from fermentation and molding.

Process for Drying of Fruits

Fruits (mature and free from insects and disease→ Washing → Peeling/ removal of outer skin→ Preparation → Pre-treatment's → Spreading on flat wooden trays → Sulphuring → Drying →Sweating → Packaging in air tight tin containers (or) polythene bags → Storage (at ambient temperature).

Pre-treatment

Lye Peeling

Dipping the fruits (grapes and dates) in 0.5 to 2.5 per cent boiling caustic soda solution for 0.5 to 2.0 minutes depending on their nature and maturity. Hot lye loosens the skin from the flesh by dissolving the pectin. The peel is then removed easily by hand. Any trace of alkali is removed by washing the fruit thoroughly in running cold water (or) dipping it for a few seconds in 0.5 per cent citric acid solution.

Sulphuring

Sulphuring is done only for fruits and not vegetables. SO_2 fumes act as a disinfectant and prevent the oxidation and darkening of fruits on exposure and thus improves their colour. This phenomenon is generally seen in sliced fruits which darken due to oxidation of the colouring matter. Sulphur fumes also act as a preservative, check the growth of molds *etc.* and prevent cut fruit pieces from fermenting while drying in the sun. Vitamins in sulphured fruits are protected but not in unsulphured ones.

The whole fruits, slices (or) pieces are exposed to the fumes of burning sulphur inside a closed chamber known as sulphur box for 30-60 min. or in small airtight rooms. Sulphur box is a closed airtight chamber of galvanized iron sheet. It is fitted in a wooden frame work having runways on both sides to hold the trays. For small scale sulphuring, a box of size of 90 x 60 x 90 cm which can hold 11 trays, each of 80 x 60 x 5 cm size is suitable. A box holding 10 trays will require burning of about 3 g of sulphur in one charge.

Sweating

Keeping dried products in boxes or bins to equalize moisture content.

Principle for Dehydrating Fruits and Vegetables

Micro organisms need moisture to grow so when the concentration of water in the food is brought down below certain level, they are unable to grow. The removal of moisture is the basic principle behind the dehydration technique. Sulphiting prevent the darkening of vegetables and improves colour.

Dehydration of Potatoes

Ingredients

- Potato - 1kg
- Potassium meta bisulphate - as required
- Potatoes were washed thoroughly.
- Sliced into thin slices.
- They were divided into four batches.
- First batch of potatoes was kept as control [without any pre treatment].
- The second batch was subjected to steam blanching for 2 minutes.
- The third batch was immersed into 1 per cent KMS solution [sulphiting] for 20 minutes
- The fourth batch is steam blanched and sulphited.
- Water was drained.
- Potato slices were kept on trays and dried at 60°C.

Dehydration of Beans

- Selected matured and firm beans.
- Washed it thoroughly and the strings were removed.
- Pods were sliced length wise.
- The first batch of beans kept as control [without any pre treatment].
- The second batch was subjected to steam blanching for 2 minutes.
- Water was drained.
- Bean slices were kept on trays and dried at 60°C.

Dehydration of Fruits (Banana)

- Optimum ripened banana were selected
- Washed in running water and peeled
- Sliced into thick slices and blanched for 3 minutes.
- Sulphited with 2 per cent KMS solution for 30 minutes.
- Pieces were drained and spread in trays.
- Dehydrated in cabinet drier at 60-65°C.

Chapter 16

Food Packaging: Importance, Types of Packaging Materials and Nutritional Labelling

Food Packaging

It is defined as the enclosure of products, items in a wrapped pouch, bag, box, cup, tray, can, tube, bottle or other container form to perform one or more of the following functions: containment, protection, preservation, communication, utility and performance.

Types/Levels of Food Packaging

Primary Food Packaging

The packaging material which comes in direct contact with product is known as primary packaging. *e.g.* laminate/wrapper, wax paper.

Secondary Food Packaging

These packaging are required basically for transportation/warehouse storage/handling. Mostly these are cardboard boxes (CBB)

Tertiary Food Packaging

Tertiary packaging is used for bulk handling, warehouse storage and transport shipping. The most common form is a palletized unit load that packs tightly.

Functions of Food Packaging

- Containment
- Protection
- Preservation
- Communication

A good package should therefore perform the following functions:

- It should provide a barrier against dirt and other contaminants thus keeping the product clean
- It should prevent losses. For example, packages should be securely closed to prevent leakage
- It should protect food against physical and chemical damage.
- The package design should provide protection and convenience in handling and transport during distribution and marketing
- It should help the customers to identify the food and instruct them how to use it correctly
- It should persuade the consumer to purchase the food.

Materials Used in Packaging	*Examples of Use*
Metals: Tin cans, aluminium cans, foil containers, foil bags, aerosols	Fruit, soft drinks, take-aways, coffee, whipped cream.
Glass: Jars, bottles	Mayonnaise, olive oil
Paper: Plain paper, waxed paper, laminated cartons, greaseproof paper, cardboard	Sugar, bread, orange juice, sausage rolls, cereal boxes
Wood: Boxes or wooden Grades	Fruits, vegetables and dried fish
Plastic: polythene, plastic cartons or plastic boxes, PET bottles, polystyrene.	Dried pasta, yoghurt, dairy spread, soft drinks, take-away food.

Advantages of different Packaging Materials

Metal: protects, easy stack and store, can be heated, lacquered to prevent reaction with food, heavy -increase transport cost, expensive, recyclable.

Glass: hygienic, protects well, rigid, variety shapes and sizes, transparent, resealable, stackable, can be heated, doesn't react with food, heavy, breakable, costly, recyclable

Paper: biodegradable, low cost, easy to open not easy to reseal, light, can be printed on, plain paper-not strong, waxed paper is moisture proof and vapour proof.

Plastic: strong, moisture proof, light, flexible, heat sealable, low cost, suitable for freezing, variety of weights, sizes, shapes, thicknesses, can be printed on, some may contaminate food when heated in microwave.

Usage of Packaging Materials

1. **Raw vegetables and fruits:** Raw vegetables may be packed loose in gunny bags or wooden/plastic crates or in lined or unlined corrugated fiber board boxes.
2. **Juices, jams, jellies and marmalades:** They should be packed in glass bottles or open top cans. Tomato juice be packed in glass or tinplate containers and hermetically sealed.
3. **Carbonated beverages**: Packed in PET bottles or glass bottles
4. **Pickles:** Packed in PET bottles or glass bottles. It should be hermetically closed to prevent spoilage.
5. **Squash and RTS** – tetra pack, PET or glass bottles are used for packaging.
6. **Bread:** bread should be wrapped in Low density poly ethylene (LDPE) coated paper or clean waxed paper, grease proof paper.
7. **Biscuits:** It should be packed in cello/LDPE, BOPP/LDPE, PET/LDPE and also paper/LDPE, foil/LDPE in such a way as to protect them from breakage, contamination, absorption or moisture and seepage of fat from the biscuits into the packing materials.
8. **Cakes:** The cakes should be wrapped or packed in clean waxed paper, grease-proof polyethylene, glassine/LDPE or any other suitable wrapper or tins.
9. **Peanut butter:** It should be packed in wide-mouthed glass jars or polystyrene tubs or any other suitable container.
10. **Flours:** The material should be packed in either LDPE coated jute bags.
11. **Oil:** The oil should be packed in suitable well closed containers. The packaging material used may be tinplate containers, glass bottles, rigid plastic containers of HDPE, food grade PVC, PET and flexible pouches made of plastic film/foil/laminate.
12. **Fat, Vanaspati:** The material should be packed in suitable sealed packages,such as, flexible packs and well closed tinplate containers.
13. **Cereal grains:** Cereal gains should be packed in new, clean jute bags orLDPE coated jute bags and raffia bags.
14. **Honey:** The packing of honey should be in hygienically clean and wide mouthed glass containers.
15. **Tea and coffee:** To maintain the flavour of tea, it should be packed in such a manner as to allow the tea to retain its freshness. Tea could be packed in flexible packaging materials or laminates such as LDPE, paper coated LDPE, PET/LDPE and BOPP/LDPE.

16. **Chocolates:** Chocolates meant for covering purposes should be packed in clean, sound and odour-free containers. Such containers may be either made of tinplate, plastic, greaseproof paper, aluminium foil or laminates made of paper/LDPE, BOPP/LDPE.
17. **Beer:** Beer may be filled in 650 ml glass bottles or in PET bottles. Bottles or cans should be packed in wooden cases or corrugated fiber board boxes.
18. **Spices and condiments:** Chillies should be packed in clean and sound jute bags or in pouches made from PET/EVA or BOPP/EVA or in suitably lined wooden cases. The material may also be packed in LDPE coated raffia bags.

Desirable Properties of Food Packaging

- Safe, non-toxic
- Hygienic
- Easy to open and reseal
- Attractive
- Economical
- Strong
- Biodegradable or recyclable
- Odourless

Controlled Atmosphere Packaging

Controlled atmosphere packaging (CAP) is the enclosure of food in a gas impermeable package and **atmosphere** (CO_2, O_2, N_2, water vapour) **inside the package is modified and** is **selectively controlled** to increase shelf life.

In this technology, the storage system consists of air tight storage chambers, O2 regulatory unit, CO_2 absorbing unit equipment for monitoring as well as controlling the chambers and composition.

For maintaining optimum level of CO_2 in the chambers, air in chambers is circulated through CO_2 scrubber frequently. CO_2 absorbing materials such as hydrated lime, calcium or potassium hydroxides are generally used in scrubber.

Refrigeration unit is employed for maintaining storage temperature. The storage life of various fruits and vegetables can be increased by 2 to 4times the normal life by employing Controlled Atmosphere storage technology. However, CA stored produce deteriorate rapidly when exposed to normal atmospheres during marketing

Modified Atmosphere Packaging

Modified Atmosphere Packaging (MAP) can be defined as the enclosure of food in package in which the **atmosphere inside the package is modified** or altered to provide an optimum atmosphere for increasing shelf life and maintaining quality of the food.

Modification of the atmosphere may be achieved either actively or passively.

Active modification - displacing the air with a desired mixture of gases

Passive modification- by foods respiration

Gases Used in MAP

The three main gases used in MAP are CO_2, O_2, and N_2, either singly or in Combination.

Carbon dioxide: Carbon dioxide is the most important gas in the MAP of foods because of its bacteriostatic and fungistatic properties. It inhibits the growth of many spoilage bacteria.

Oxygen: Oxygen promotes several types of deteriorative reactions in foods including fat oxidation, browning reactions and pigment oxidation. Most of the common spoilage bacteria and fungi require O2 for growth.

Nitrogen: Nitrogen is an inert gas with no odor or taste. It has a lower density than air and a low solubility in water and other food constituents, making it a useful filler gas.

Advantages and Disadvantages of Modified Atmosphere Packaging

Advantages

1. Shelf life of the product will be increased by MAP
2. Reduced economic losses due to longer shelf life.
3. Provides a high quality product.
4. Centralized packaging and portion control.
5. Improved presentation – clear view of product and all –around visibility.
6. Little or no need for chemical preservatives.
7. Sealed packages are barriers against product recontamination and drip from Package.
8. Odour less and convenient packages.

Disadvantages

1. Added costs for gases, packaging materials and machinery.
2. Temperature control necessary.
3. Different gas formulations for each product type.
4. Special equipment and training required.
5. Increased pack volume adversely affects transport costs and retail display space.

6. Loss benefits once the pack is opened or leaks.
7. CO_2 dissolving into the food could lead to pack collapse and increased drip.

In controlled atmosphere storage (CAS), the gas composition inside a food storage room continually monitored and adjusted to maintain the optimum concentration within quite close tolerances. In contrast, the less common modified atmosphere storage (MAS) typically involves some initial modification of the atmospheric composition in an airtight storage room, which changes further with time as a result of the respiratory activity of the fresh food and the growth microorganisms.

Active Packaging

The terms active packaging, intelligent packaging, and smart packaging refer to packaging systems used with foods, pharmaceuticals, and several other types of products. They help extend shelf life, monitor freshness, display information on quality, improve safety, and improve convenience. Some of the techniques used in active packaging are

- Moisture control
- Corrosion
- Oxygen control
- Atmosphere
- Temperature monitor
- Controlling package temperatures
- Dispensing
- Security
- Microwave packaging
- Shock and vibration

Food Label

A food label can be defined as a panel found on a package of food which contains a variety of information about the nutritional value of the food item. There are many pieces of information which are standard on most food labels, including serving size, number of calories, grams of fat, included nutrients, and a list of ingredients. This information helps people who are trying to restrict their intake of fat, sodium, sugar, or other ingredients, or those individuals who are trying to get enough of the healthy nutrients such as calcium or Vitamin C.

Food label is any tag, brand, mark, pictorial or other descriptive matter, written, printed, stencilled, marked, embossed or impressed on, or attached to, a container of food. Food labelling includes any written, printed or graphic matter that is present on the label, accompanies the food, or is displayed near the food, including that for the purpose of promoting its sale or disposal. The label may either be engraved or attached on to the packaging of the product.

Nutritional Labelling

Food which provide nutrition information on a voluntary basis on the lable have been required to provide to provide nutrition information in the prescribed format.

Nutritional labelling include the following information in prescribed manner

A. The energy value

B. The amount of fat, saturates, carbohydrate, sugars, protein, and salt

The content of the mandatory nutrition declaration may be supplemented with an indication of the amounts of one or more of the following

a. Monosacharides

b. Polysaccharides

c. Polyols

d. Starch

e. fibre

Food Labelling

Functions of Food Labelling

- ✰ Inform consumer about the nature of pre-packed food
- ✰ Identify product
- ✰ Help sell product

The Information Supplied must Cover

- ✰ The name of food
- ✰ The weight/volume
- ✰ The list of ingredients in descending order of weight
- ✰ How the food should be stored
- ✰ The date when the food should be eaten- date code
- ✰ Genetically modified ingredients and allergens
- ✰ The name and address of manufacturer

Quick Easy Reference for Consumers

- ✰ Foods with green lights, healthier choice.
- ✰ An amber means neither high or low so you can eat these foods often.
- ✰ Red lights mean that these foods are high in fat, saturated, salt, or sugars and we should eat less of these.

Chapter

17

Food Adulteration

For good health, we must have good food. By good food, we mean nutritionally balanced, wholesome and safe food. Adulteration in food products is a part of under nutrition. Adulteration leads to loss of nutrients in food products and sometimes may prove very harmful for the body. Moreover, the consumer is cheated not only for his money by adulteration but also in quantity and quality as well. Adulteration of food stuffs is commonly practised in India by the trade. The consumers like to get maximum quantity for a price as low as possible. The sellers must meet the needs of the buyers, to be able to exist. This is a vicious cycle. When the price of food production is higher than the price which the consumer is prepared to pay, seller is compelled to supply food product of inferior quality. Thus, adulteration occurs.

Food adulteration can be defined as the intentional addition or substitution or abstraction of substance which adversely affect the nature, substance and quality of food products. It can be at the time of harvesting, storage, processing, transportation or distribution.

Types of Adulteration

Adulteration may be intentional or unintentional. The former is a willful act on the part of the adulterator intended to increase the margin of profit. Incidental contamination is usually due to ignorance, negligence or lack of proper facilities.

A. **Intentional adulterants:** Intentional adulterants are sand, marble chips, stones mud, chalk powder, water, mineral oil and coaltar, dyes. These adulterants cause harmful effects on the body.

B. **Incidental adulterants:** Raw foods such as meat, fish, milk and vegetables grown on sewage are likely to be contaminated with harmful microorganisms. These are generally destroyed during cooking or processing of food. Some of the microorganisms may survive due to inadequate heat processing. Further, some of the foods, if consumed in the raw state, may cause food poisoning. Recent studies have shown that food grains legumes and oil seeds when stored in humid atmosphere are infected by pathogenic fungus which can cause serious illness.

To check the growing tendency of food adulteration, Government of India had introduced "Prevention of Food Adulteration Act" (PFA) in 1954. This Act is based on international standards, keeping in mind the Indian conditions. This Act came into force on June 1, 1955.

Adulteration is defined as the process by which the quality or the nature of a given substance is reduced through the addition of a foreign or an inferior substance and the removal of a vital element. A good example for the first one is addition of water to milk and that for the second is removal of fat from milk. Adulteration of food may endanger due to either addition of a deleterious substance of the removal of a vital component.

Adulterant is any material which is employed or which could be employed for the purpose adulteration is defined as an adulterant.

Identification of Common Food Adulterants

	Food Stuff	*Adulterant*	*Test*
1.	Milk, curd, ghee, Butter	Starch	Add a drop iodine solution to a little of sample. Blue colour indicates added starch in any form. (Iodine solution is prepared by dissolving 2.5g of iodine crystals and 3g of potassium iodide in sufficient water to make the volume 100ml.
2.	Turmeric (whole and powdered and mixed spices powdered)	Metanil yellow	Shake sample with some water. Dilute water till is almost colourless and then add a few drops of Conc.hydrochloric acid. Magenta red colour seen if dye present.
3.	Wheat flour, suji, bengalgram flour	Chalk powder	Treat sample with hot dilute hydrochloric acid. Bubbling of gas indicates carbon dioxide from chalk or other carbonates.

	Food Stuff	Adulterant	Test
4.	Tea	Artificially coloured foreign matter or exhausted tea leaves	Sprinkle the tea on a sheet of wet white paper. Pink or red spots appearing on the paper indicate added colour.
5.	Ghee or butter	Vanaspathi	Dissolve a pinch of cane sugar in 10 ml concentrated hydrochloric acid taken in a glass stoppered test tube. Add 10 ml of the melted ghee, stopper the bottle and shake vigorously for two minutes. Allow to stand still 2 layers separate. If the lower layer turns pink or red, the ghee contains vanaspathi.
6.	Edible oils	Argemone Oil	Add concentrated nitric acid to the sample and shake carefulty. Red to reddish brown colour in acid layer would indicate the presence of argemone oil
		Mineral Oil	Take two ml of edible oil and add an equal quantity of N/2 alcoholic potash. Heat in boiling water both for 15 minutes and add 10 ml oi water. Any turbidity shows the presence of mineral oil.
		Castor Oil	Dissolve some oil in petroleum ether in a test tube and cool in ice salt mixture. Presence of turbidity within 5 minutes indicates the presence of oil. * This test is not for minute traces
7.	Wheat, bajra and other food grains	Ergot (a fungus containing a poisonous substance)	a. Purple black longer size grains in bajra show the presence of ergots, b. Put some grains in a glass containing 20 per cent salt solution. Ergot floats over the surface while sound grains settle down.
8.	Sugar	Chalk powder	Dissolve in a glass of water, chalk will settle down at the bottom.

	Food Stuff	*Adulterant*	*Test*
9.	Dhals	Kesari dhal	Add 50 ml of dilute Hydrochloric acid to dal and keep on simmering
10.	Tea leaves	Exhausted tea or black or bengal gram dal husk with colour	a. Tea leaves sprinkled on wet filter paper would immediately release added colour.b. Spread a little slaked lime on white porcelain tile or glass plate. Sprinkle a little tea dust on the lime. Red orange or other shades of colour spreading on the lime will show the presence of coal tar dye, in the case of genuine tea, there will be only a slight greenish yellow colour due to chlorophyll which appears after sometime.
11	Chilli Powder	Stones Artificial colour	Any grittiness that may be felt on tapping the sediment at the bottom of glass confirms the presence of brick powder or sand. Smooth white residue at the bottom indicates the presence of soapstone. Water soluble artificial dye can be detected by sprinkling a small quantity of chilli or turmeric powder on the surface of water contained in a glass tumbler. The soluble dye will immediately start discending in colour streaks.
12	Jaggery Powder	Chalk powder	Add few drops of Hcl. Effervescence indicates adulteration. Stir a spoonful sample of sugar in a glass of water. The chalk settles down.
13	Rawa	Iron filings	By moving a magnet through it iron fillings can be separated.
14	Common Salt	White powdered stone, chalk	Stir a spoonful of simple of salt in a glass of water. The presence of chalk will make the solution white and other insoluble impurities will settle down.
15	Honey	Molasses (sugar and water)	A cotton wick dipped in pure honey when lighted with a match stick burns. 1f adulterated the presence of water will not allow the honey to burn. If does it will produce a crackling sound.

	Food Stuff	Adulterant	Test
16	Coffee	Chicory	Gently sprinkle the coffee powder sample on the surface of water in a glass. The coffee floats over the water but chicory begins to sink down within a few seconds. The falling chicory powder particles leave behind them a trail of colour due to large amount of caramel they contain.
		Tamarind or date- seed powder.	Sprinkle the suspected coffee powder on white blotting paper and spray over it 1 per cent sodium carbonate solution. Tamarind and date- seed powder will, if present, stain blotting paper red.
17	Khoa	Starch	Add tincture of iodine. Indication of blue colour shows the presence of starch.

Harmful Effects of Adulterants

In the earlier times, people used to prepare different eatables at home so that it is fresh, clean and healthy. But today everyone is so busy that they do not have time to cook food at home and rely on instantly made items. These have different kinds of preservatives that are used to keep them in good condition for a long time. But these are really harmful and have an adverse effect on our health. Today almost all the food items that you will get in the market have some or the other kind of adulterant which brings down the nutritional value and causes different types of health hazards. Let's know more about adulteration and how it is harmful to your health.

1. Metanil Yellow

Metanil yellow is a food colorant used extensively in various foods. It is also used as an adulterant in different spices especially turmeric and regular Indian foods like pulses. Metanil yellow belongs to "non-permitted" category of food colour. Continued consumption of this toxic food color leads to adverse life-threatening effects in humans. Consumption of metanil yellow in food may affect our nervous system and may cause brain damage. Metanil yellow consumed through food enters our digestive system directly. Studies reveal that metanil yellow causes gastrotoxicity, hepatotoxicity, and damages the intestine.

2. Argemone Oil

Consumption of adulterated mustard oil (*Brassica nigra*) with argemone oil (*Argemone mexicana*) even for a short duration leads to a clinical condition referred as epidemic dropsy. In humans, argemone oil contained in adulterated mustard oil causes oxidative stress and death of red blood cells via met-hemoglobin formation by altering pyridine nucleotide(s) and glutathione redox potential. Argemone oil contamination poses a serious threat to human health.

3. Kesari Dal

Excessive consumption of kesari dal produces Lathyrism (form of crippling, paralysis of both lower limbs) mainly in boys and men in the age group of 5 to 45 years. This disease manifests itself in 2-4 months if the diet consists of 40 per cent or more of kesari dal.

- The disease starts with the stiffness of knee joints and legs, with pain around the knee and ankle joints.
- Within 10-30 days of the onset of symptoms, paralysis of the lower limbs sets in.
- The patient gradually becomes crippled as his knees are bent and stiff.

Chapter 18

Food Laws, Regulations and Quality Control Standards

Every country needs laws to encourage the production of safe and wholesome foods, and to prohibit the sale of foods that are unsafe or fraudulent.

Prevention of Food Adulteration Act, 1954

A basic statute (Prevention of Food Adulteration Act (PFA) of 1954 and the PFA Rules of 1955, as amended) protects India against impure, unsafe, and fraudulently labelled foods. The Act was promulgated by Parliament in 1954 to make provision for the prevention of adulteration of food, along with the Prevention of Food Adulteration Rules, 1955 which was incorporated in 1955 as an extension to the Act.

The PFA Act covers food standards, general procedures for sampling, analysis of food, powers of authorized officers, nature of penalties and other parameters related to food. It deals with parameters relating to food additives, preservative, colouring matters, packing and labelling of foods, prohibition and regulations of sales *etc.*

Amendments in PFA rules are incorporated with the recommendation made by the Central Committee of Food Standards (CCFS) which has been setup by Central Government under the Ministry of Health and Family Welfare comprising members from different regions of the country. The provisions of PFA Act and Rules are implemented by State Government and local bodies as provided in the rules.

Prevention of Food Adulteration Act, 1954 will be repealed from the date to be notified by the Central Government as per the Food Safety and Standards Act,2006. Till that date new standards are specified, the requirement and other provisions of the PFA Act, 1954 and Rules, 1955 shall continue to be in force as a transitory provision for food standards.

Fruit Products Order (FPO), 1955

FPO constituted under Essential Commodities Act it is mandatory for all manufacturers of Fruit and Vegetable Products to obtain a licence under FPO. The act is implemented by the Food Safety and Standards Authority of India through Directorate of Fruit and Veg. Processing at its Regional Offices.

The Fruit Product Order (FPO) lays down statutory minimum standards in respect of the quality of various fruits and vegetable products and processing facilities at manufacture, storage and sale. The Agricultural marketing Advisor is authorized by law to issue a license for manufacturing fruits and vegetable products, after due inspection of the factory for hygiene, sanitation and quality of formulation. Periodic inspection by Government inspectors in establishments is carried out to ensure conformity of standards by processors. Licenser is empowered to put the FPO specification mark on the product. The products covered in FPO include, fruit juice, pulp concentrate, squashes, cordials, crush, fruit syrups, nectar, aerated water containing fruit juice or pulp and read to serve beverages *etc.*, Depending on their quality the products are grade in four categories as ordinary, fair, good and special. The FPO specifications cover list of constituents, a method of presentation permissible colours in the preparation and also minimum quality requirement of the product. An expert committee known as the Central Food Product Advisory Committee deals with all matters relating to the FPO.

FPO also lays down specific requirements in regard to the following:

- Containers and labelling requirement
- Limits of poisonous metals in fruit products
- List of permissible harmless food colors
- Limits for permitted preservatives in fruit products
- Other permitted additives

The Meat Food Products Order, 1973

The main objectives of the MFPO, 1973 are to regulate production and sale of meat food products through licensing of manufacturers, enforce sanitary and hygienic conditions prescribed for production of wholesome meat food products, exercise strict quality control at all stages of production of meat food products, fish products including chilled poultry *etc.*

Under the provision of MFPO all manufacturers of meat food products engaged in the business of manufacturing, packing, repacking, relabeling meat food products meant for sale are licensed but excluding those manufacturers who manufactures such products for consumption on the spot like a restaurant, hotel, boarding house, snack bar, eating house or any other similar establishment.

The Vegetable Oil Products (Control) Order, 1947

The Vegetable Oil Products industry is regulated by this Order through the Directorate of Vanaspati, Vegetable Oils and Fats, Department of Food, Public Distribution, Ministry of Consumer Affairs, and Food and Public Distribution. The earlier two Orders – Vegetable Oil Products (Control) Order, 1947 and Vegetable Oil Products (Standards of Quality) Order, 1975 have been replaced by a single Order called "Vegetable Oil Products (Regulation) Order, 1998 for proper regulation of manufacture, distribution and sale of Vegetable Oil Products.

Salient Features of the Order:

- The procedure of Registration has been simplified.
- The Standards of quality prescribed under the Schedule have been tightened.
- The requirement where which are vogue and non measurable and thus open to arbitrary interpretation have been done away with.
- Consumers' protection through quality assured.

The Edible Oils Packaging (Regulation) Order, 1998

The Department of Sugar and Edible Oil under Ministry of food and Consumer Affairs regulate this-order.

The Solvent Extracted Oil, De oiled Meal, and Edible Flour (Control) Order, 1967

It applies for oils and fats and deals with the licensing, manufacture, distribution and trade of solvent extracted edible oils and quality thereof and operated by Directorate of Vanaspati, Vegetable Oil and fats.

Milk and Milk Products Order, 1992

The production, distribution and supply of milk products are controlled by the Milk and Milk Products Order, 1992. The order sets sanitary requirements for dairies, machinery, and premises, and includes quality control, certification, packing, marking and labeling standards for milk and milk products. Standards specified in the order also apply to imported products. The Department of Animal Husbandry, Dairying and Fisheries at the Ministry of Agriculture is the regulatory authority.

Livestock Importation Act, 1898

India has established procedures for the importation of livestock and associated products under the Livestock Importation Act, 1898. Under the regulations, the import of meat products, eggs and egg powder and milk products require a sanitary import permit from the Department of Animal Husbandry, Dairying and Fisheries at the Ministry of Agriculture. A detailed import risk analysis is carried out, taking into account the disease situation prevailing in the exporting country compared with the disease situation in India.

Essential Commodities Act, 1955

The main objective of the Act is to regulate the manufacture, commerce, and distribution of essential commodities, including food. A number of Control Orders have been promulgated under the provisions of this Act. These are:

Standards of Weights and Measures Act, 1976 and the Standards of Weights and Measures (Packaged Commodities) Rules, 1977

The Act governs sale of packaged commodities and provides for mandatory registration of all packaged products in the country.

Consumer Protection Act, 1986

The Act provides for constitution of District Forum/State/National Commission for settlement of disputes between the seller/service provider and the consumer.

The Infant Milk Substitutes, Feeding Bottles and Infant Foods (Regulation of Production, Supply and Distribution) Act, 1992 and Rules 1993

This Act aims at promoting breast feeding and ensuring proper use of infant milk substitutes and infant food.

The Insecticide Act, 1968

The Act envisages safe use of insecticides so as to ensure that the leftover chemical residues do not pose any health hazard.

Export (Quality Control and Inspection) Act, 1963

The Act aims at facilitating export trade through quality control and inspection before the products are sold to international buyers.

Environment Protection Act, 1986

This Act incorporates rules for the manufacture, use, import and storage of hazardous microorganisms/substances/cells used as foodstuff.

Pollution Control (Ministry of Environment and Forests): A no-objection certificate from the respective State Pollution Control Board is essential for all dairy plants.

Harmonization of Food Laws

Government of India took initiation for the harmonization of various laws and acts related to food Safety and the regulatory standards for food safety. As a result of many meetings and deliberations, GOI passed a bill in parliament of India for the Food Safety and StandardsBill.2006 for Harmonization of food standards and single platform for food safety regulatory standards.

Food Safety and Standard Act 2006 (FSSAI)

The act has been enacted to consolidate the laws related to food and to establish the Food safety and standards Authority of India for laying down science based standards for articles of food and to regulate their manufacture, storage distribution sale and import to ensure availability of safe and whole some food for human consumption.

Salient Features of FSSAI Act, 2006

Some of the salient features of the Act are:

- The Act integrates eight different food related statutes
  - Prevention of Food Adulteration Act, 1954,
  - Fruit Products Order, 1955,
  - Meat Food Products Order, 1973,
  - Vegetable Oil Products (Control) Order, 1947,
  - Edible Oils Packaging (Regulation)Order 1988,
  - Solvent Extracted Oil, De- Oiled Meal an and Edible Flour (Control) Order, 1967,
  - Milk and Milk Products Order, 1992.
- FSSAI as a **single reference poin**t for all matters relating to Food Safety and Standards, Regulations and Enforcement.
- Integrated **response to strategic issues** like Novel foods, Health Foods, Nutraceuticals, GM foods, international trade *etc.*
- Adequate **information dissemination** on food to enable consumer to make informed choices.
- **Graded penalty** depending upon the gravity of offences.
- Adequate **representation of government**, industry organizations, consumers, farmers, technical experts, retailers *etc.*
- **Enforcement of the legislation** by the State Governments/UTs through the state Commissioner for Food Safety, his officers and Panchayati Raj/ Municipal bodies.

AGMARK

The word Agmark is derived from Agricultural Marketing. The DMI under the Department of Agriculture and Co-operation in the Ministry of Agriculture enforces the Agricultural Products (Grading and Marketing) Act 1937. Under this Act Grade standards are prescribed for agricultural and allied commodities. Agmark grading means grading of an article in accordance with grade/standards prescribed under the provisions of the act.

The quality of the product is determined based on the size, variety, weight, color, moisture, fat content and other factors. It covers quality assurances of unprocessed, semi processed and processed agricultural commodities. The grades incorporated are grades 1, 2, 3 and 4 or special, good, fair and ordinary.

Salient Features of Agmark Standard

- Quality standards for Agricultural commodities are framed Food safety factors are being incorporated in the standards to compete in world trade.
- Standards are being harmonized with international standards
- Certified products through 23 laboratories and 43 offices spread all over the country

Products available under AGMARK are pulses, wheat products, vegetable oils, ground spices, whole spices, milk products, honey, compounded asafoetida, rice, tapioca sago, seedless tamarind and gram flour; grading of these commodities is voluntary. On the other hand grading of commodities like tobacco, walnut, spices, basmati rice, essential oils, onion, potatoes are meant for export is compulsory under AGMARK.

The Directorate of marketing and inspection of central government has 21 laboratories and 50 sub offices spread all over the country. The central AGMARK laboratory at Nagpur continuously carries out research and development works in this field.

Bureau of Indian Standards (BIS)

BIS is the National Standards Organization established as Society in 1947 as Indian Standards Institution and subsequently made its statutory body as BIS under Bureau of Indian Standards Act 1986. It revoked Indian Standards Institutions (Certification Marks) Act 1952 but incorporated all its provisions. The Bureau is a body corporate and responsible for laying down policy guidelines for BIS. It comprises of members Representing Industry, Consumer Organizations, Scientific and Research Institutions, Professional/technical institutes, Central Ministries; State Government and Members of Parliament. The act is mandatory for milk powders, sweetened condensed milk, infant formula *etc.* This act generally covers hygienic conditions of manufacture, raw material quality and safety. It also ensures the quality to the consumers by certification.

Salient Features of the BIS

- Standard Formulation,
- Certification : Product Quality Management System, Eco Mark, Environment Management System, Hazard Analysis' and Critical Control Points,
- Laboratory: Testing, Calibration and Management,
- Standards Promotion,
- Consumer Affairs and
- Awareness and Training Programs.

There are 14 different technical departments engaged in formulation of the standards. So far 17000 standards have been formulated in different technological areas depending upon the national priority. These standards are evolved through the consensus from different sectors like industry, consumers, testing and laboratory experts and committees/sub-committees of Govt organization. The standards are reviewed time to time and continuously updated to match the technological changes. The BIS has formulated 1133 standards pertaining to food products.

Codex Alimentarius Commission (CAC)

The term Codex Alimentarius is taken from Latin and means food code. The Codex Alimentarius Commission develops food standards, guidelines and related texts such as codes of practice under the Joint FAO/WHO Food Standards Programme. It is also called Codex

Powers and Functions of the Bureau

- Establishment, Publication and Promotion of Indian Standards
- Procedure for Establishment of Indian Standards
- Grant of License
- Inspection
- STOP Marking

(Hormonized International Standards due to involvement of both FAO and WHO. About 170 countries were member of the commission. The purpose of this program is to protect the health of consumers and to ensure fair practice in the food trade; to promote coordination of all food standards work undertaken by international governmental and non-governmental organizations; to determine priorities and initiate and guide the preparation of draft standards through and with the aid of appropriate organizations; to finalize standards and after acceptance by Governments, publish them in a Codex Alimentarius either regional or worldwide standards. It brings together all the interested parties *viz.* scientists, technical experts, governments, consumers and industry representatives to help develop standards for food manufacturing and trade. These standards, guidelines and recommendations are recognized worldwide for their vital role in protecting the consumer and facilitating international trade. As Codex Alimentarius represent a consensus of food and trade experts from around the world, these standards are more and nl0re being used in international trade negotiations and also for setting of disputes by WTO. The Codex contract Point in India is the Directorate General of Health Services (DGHS) in the Ministry of Health; however, the Ministry of Food processing Industries is closely associated with the activities of Codex Alimentarius.

Salient Features of Codex Alimentarius

- Protecting health of the consumers and ensuring fair trade practices
- Promoting coordination of all food standards work undertaken by international governmental and non-governmental organizations

- ☆ Determining priorities initiating and guiding the preparation of draft standards.
- ☆ Finalizing standards
- ☆ Amending published standards
- ☆ Submission of a proposal for a standard
- ☆ A decision by the Commission or the executive committee
- ☆ Preparation of a proposed draft standard by subsidiary body
- ☆ Adoption of standard by the Commission
- ☆ Addition of Codex Standard in the Codex Alimentarius

World Trade Organisation (WTO)

WTO was established in 1995. The main objective of WTO is to help trade flow smoothly, freely, fairly and predictably, by administering trade agreements, settling trade disputes, assisting countries in trade policy issues. The WTO Agreement covers goods, services and intellectual property. In order to enforce adoption and implementation of standards, there is a need for a strong Food Control System. An effective food control system must consist of – (i) Food Inspection and (ii) Analytical capability. Food Inspection: Conformity of products to standards is verified through inspection. This will ensure that all foods are produced, handled, processed, stored and distributed in compliance with regulations and legislation. Government/ Municipal authorities appoint food inspectors to investigate the status of quality conformity to standards in their laboratories. Analytical capability: There is need for well-equipped, state-of-the-art accredited laboratories to carry out analysis of food. Further, well- trained personnel having knowledge of principles of laboratory management and physical, chemical and microbiological analysis of food, test foods and food products are also required. A broad range of analytical capabilities is required for detecting food contaminants, environmental chemicals, biotoxins, pathogenic bacteria, food-borne viruses and parasites.

Food Safety Management Systems

Over the years, issues related to food safety and quality have gone beyond just the avoidance of food-borne pathogens, chemical toxicants and other hazards. A food hazard can enter/come into the food at any stage of the food chain, therefore, adequate control throughout the food chain is essential.

Food safety and quality can be ensured through:

- ☆ Good Manufacturing Practices (GMP)
- ☆ Good Handling Practices (GHP)
- ☆ Hazard Analysis Critical Control Points (HACCP)

Good Manufacturing Practices (GMP) are a part of quality assurance to ensure that manufacturers/processors take proactive steps to ensure that their products are safe. It enables to minimise or eliminate contamination and false labelling, thereby protecting the consumer from being misled and helping in purchasing products that are not harmful. GMP is a good business tool that helps to refine compliance and performance by the manufacturers/producers.

Good Handling Practices (GHP) indicate a comprehensive approach from the farm to the store or consumer, in order to identify potential sources of risk and indicates what steps and procedures are taken to minimise the risk of contamination. It ensures that all persons who handle food have good hygiene practices.

Hazard Analysis Critical Control Point (HACCP)

The Hazard Analysis Critical Control Point (HACCP) is a scientific, rational and systematic approach to identify, assess and control hazards during production, processing, manufacturing and use of food. It ensures safety of the food. A systematic approach to the identification, evaluation, and control of food safety hazards based on seven principles.

History of HACCP

The application of HACCP to food production was pioneered by the **Pillsbury Company** with the cooperation and participation of the **National Aeronautic and Space Administration (NASA), Natick Laboratories of the US Army, and the US Air Force Space Laboratory Project Group**. In the Indian food industry, HACCP is not compulsory practice but due to increase in the cases of food poisoning in the country, HACCP has been recognized internationally as a science-based logical tool for management of food safety. India as signatory to WTO, TBT and SPS agreements along with TQM and ISO is committed to follow WTO regulations whose priority is to protect health of consumer. Our food industry thus needs to install HACCP to avoid any rejection of foods shipped to international market.

Seven Principles of HACCP

1. **Conduct hazard analysis:** In this first step, the team assesses hazards associated with growing, harvesting, raw materials and ingredients, processing, manufacturing, distribution, marketing, preparation, and consumption of the food. They identify all significant hazards (biological, chemical, and physical) that need to be controlled to assure food safety throughout each step in the process.
2. **Determine critical control points:** After identifying the significant hazards, the team establishes preventative measures to control the identified hazards. The team identifies areas or points in the flow of a food product with their critical limits that must be met to control the identified hazards. These are called critical control points (CCPs). **A Critical Control Point (CCP) is a point, step, or procedure in a food process at which control can be applied and as a result, a food safety hazard can be**

prevented, eliminated, or reduced to an acceptable level. If loss of control occurs at a CCP, it likely will lead to an unacceptable health risk.

3. **Establishment of specification for critical limits:** At each CCP, teams define boundaries or limits of safety to assure that the CCP is in control. They establish upper limits for CCPs. CCP limits are usually based on time, temperature, pH, and moisture content of a food.
4. **Development of monitoring and testing system to control critical point:** CCP and CCP limits are only effective if they are monitored during food processing. Monitoring ensures that the process is in control. Testing procedures have to be developed to ensure that each critical control point is consistently monitored and the process is under control.
5. **Establishment of corrective actions when particular CCP is not under control:** Whenever food companies note a deviation in the critical limits for a CCP, they must correct the deviation. Corrective action may include changing the process, reprocessing, or discarding the product. Corrective actions are intended to ensure that no product injurious to health.
6. **Establish record keeping procedures:** Food companies must keep records of the results of monitoring critical control points. These records are the only proof for a company that process is in control and that they are complying with the HACCP plan. Depending on the commodity, records should be accessible for one to three years. Thus, a system must establish documentation of procedures and records for all aspects of the HACCP programme and give evidence of its functioning base on all data obtained from testing and analysis, deviation or correction actions.
7. **Verification of HACCP system to confirm efficacy:** The procedure need to be established to verify and confirms that operating HACCP system is working effectively. Verification ensures that HACCP plan is adequate and verification includes such activities as review of HACCP plans. They may request an internal or external food safety audit to verify that the HACCP plan is working.

Benefits of HACCP

There are numerous benefits for the food industry while applying HACCP system as a management tool for food safety control. Some of the important benefits are as follows:

1. Application of the HACCP concept is the cost effective approach to food safety.
2. Application of the HACCP concept is enough flexible.
3. Helps to maintain the global food quality and safety standards.
4. The HACCP approach is a systematic approach for all aspects of food safety and can be applied to all stages of the food chain, including raw

materials, growth, harvesting, purchase, production, distribution, and storage to final product use.

5. Provides scientifically sound base for protections of a hazard from reaching the end consumer products.
6. HACCP systems can promote international trade by increasing confidence in food safety.
7. HACCP system can facilitate the design and construction of new food processing facilities and equipment.
8. The HACCP system can be readily integrated in to quality management systems like TQM and ISO 9000 *etc.*
9. HACCP system focuses resources mainly on those parts of the process which are critical for assuring safe products.
10. HACCP system can reduce product losses due to spoilage.

Food and Drug Administration

The **Food and Drug Administration** (**FDA** or **USFDA**) is a federal agency of the United States Department of Health and Human Services, one of the United States federal executive departments. The FDA is responsible for protecting and promoting public health through the control and supervision of food safety, tobacco products, dietary supplements, prescription and over-the-counter pharmaceutical drugs (medications), vaccines, biopharmaceuticals, blood transfusions, medical devices, electromagnetic radiation emitting devices (ERED), cosmetics, animal foods and feed and veterinary products.

The FDA was empowered by the United States Congress to enforce the Federal Food, Drug, and Cosmetic Act, which serves as the primary focus for the Agency; the FDA also enforces other laws, notably Section 361 of the Public Health Service Act and associated regulations, many of which are not directly related to food or drugs. These include regulating lasers, cellular phones, condoms and control of disease on products ranging from certain household pets to sperm donation for assisted reproduction.

The FDA is led by the Commissioner of Food and Drugs, appointed by the President with the advice and consent of the Senate. The Commissioner reports to the Secretary of Health and Human Services. Scott Gottlieb, M.D. is the current commissioner, who took over in May 2017.

The FDA has its headquarters in unincorporated White Oak, Maryland. The agency also has 223 field offices and 13 laboratories located throughout the 50 states, the United States Virgin Islands, and Puerto Rico. In 2008, the FDA began to post employees to foreign countries, including China, India, Costa Rica, Chile, Belgium, and the United Kingdom

International Standards Organization (ISO)

The product sell based on the product quality as perceived by the customer is the major factor for sustained what makes a sales of a product. The method of quality control consisted largely of physical inspection of the end product against the product specification. However, technical specifications may not by themselves guarantee, that a customer's requirements will be met, if there happens to be any deficiency in the specifications or in the organizational system, to design and produce the products, or service. Consequently, this has led to the development of quality system standards and guidelines that complement relevant product, or service, requirements given in the technical specifications.

ISO 22000:2005

It specifies requirements for a food safety management system where an organization in the food chain needs to demonstrate its ability to control food safety hazards in order to ensure that food is safe at\ the time of human consumption. It is applicable to all organizations, regardless of size, which are involved in any aspect of the food chain and want to implement systems that consistently provide safe products. The means of meeting any requirements of ISO 22000:2005 can be accomplished through the use of internal and/or external resources.

ISO 22000:2005 specifies following requirements to enable organization:

- To plan, implement, operate, maintain and update a food safety management system aimed at providing products that, according to their intended use, are safe for the consumer
- To demonstrate compliance with applicable statutory and regulatory food safety requirements
- To evaluate and assess customer requirements and demonstrate conformity with those mutually agreed customer requirements that relate to food safety, in order to enhance customer satisfaction
- To effectively communicate food safety issues to their suppliers, customers and relevant interested parties in the food chain
- To ensure that the organization conforms to its stated food safety policy
- To demonstrate such conformity to relevant interested parties and
- To seek certification or registration of its food safety management system by an external organization, or make a self-assessment or self-declaration of conformity to ISO 22000:2005.

References

Adams M R and Moses M G. 1995. Food Microbiology. 1st edition, New Age International Limited, New Delhi.

Amihud Kramer and Bernard A – Twigg, Quality control for the Food Industry, AVI Publishing and Co, Connecticut, 1973.

Antia, F. P and Abraham, P. 2005. Clinical Nutrition and Dietetics, 5th edition, Oxford University Press, India, 540p.

Bakhru, H. K. 2012. Vitamins that Heal Natural Immunity for Better Health, Orient paperbacks publishing, New Delhi, India, 210p.

Bamji, M. S. Krishnaswamy, K and Brahman, G.N.V. 2016. Textbook of Human Nutrition, 4th edition, Oxford and IBH Publishing Company Limited, New Delhi. 505p.

Barrett D M and Lloyd. B. 2011. Advanced preservation methods and nutrient retention in fruits and vegetables, Journal Science Food Agriculture. 92: 7-22.

Bennion, M. 1980, Seventh Edition, Introductory Foods, Macmillan Publishing Co, NewYork. 587p.

Bhutani R. C. 2003. Fruit and Vegetable Preservation. Biotech Books, Daya Publishing House, New Delhi. 485p.

Chintapalli Vidhya and Digumarti Bhaskara Rao. 2004. 'A text book of nutrition', Discovery Publishing House, New Delhi, 437p.

Clarke, D and Elizabeth, H. 1986. Food Facts A Study of Food and Nutrition, MacMillan Education, New Delhi 320p.

Codex Alimentarius. 1998. General standard for contaminants and toxins in foods. Annex IVB, Waningen, The Netherlands.

Cody, V. Middleton, E and Karborne J. B. 1988. Plant Flavonoids in Biology and Medicine II: Biochemical, Cellular, and Medicinal Properties. Alan R Liss, Inc, New York, 213: 319–331.

Davidson S, Passmore R and Eastwood M A. 1990. Human Nutrition and Dietetics, Churchill Livingstone, 735p.

Deb A C. 1992. Fundamentals of Bio chemistry, 5th Edition, New Central Book Agency private limited.

Desrosier, N.W. and Desrosier, J.N. 1987. The Technology of Food Preservation. 4th ed. CBS Publishers and Distributors, New Delhi.581p.

Devlin T M. 1986. Text Book of Biochemistry with Clinical corrections, 2nd Edition, John Wiley and sons, Butler and Tanner, London.

Eugene L G and Richard L L. 1979.Stastical quality control, Mc. Graw Hill International Publishers.

Giridhari, Lal, Siddappa, G.S., and Tandon, G.L. 2011. Preservation of Fruits and Vegetables, Publication and Information Division, ICAR. New Delhi. 488p.

Gopalan C, Ramasastri, B. V. and Balasubramanian, S. C. 2010. Nutritive value of Indian Foods, National Institute of Nutrition, Hyderabad, 58p.

Hasler C M. 1998. Functional foods: their role in disease prevention and health promotion. Food Technology, 52: 52-70.

Herschdoerfer, S.M, 1984. Quality Control in the Food Industry. Academic press, 1: 12-18.

Jay James M. 2000. Modern Food Microbiology, 2nd edition, CBS Publisher.

Jood, S. and Khetarpaul, N. 2001. Food preservation. Agrotech Publishing Academy, Udaipur. 264p.

Joshi S. 2010. Nutrition and Dietetics, 3rd edition, Tata McGraw-Hill Education, New Delhi, 605p.

Kalia, M. and Sood, S. 2006. Food Preservation and Processing. Kalyani Publishers, Ludhiana. 555p.

Khetarpaul, N. 2005. Food processing and preservation. Daya Publishing House, Delhi. 555p.

Kulshrestha, S.K. 1994. Food preservation. Vikas Publishing House Pvt. Ltd., New Delhi. 98p.

Lampe J W. 1999. Health effects of vegetables and fruits: assessing mechanisms of action in human experimental studies, Annual Journal Clinical Nutrition, 70: 47-90.

Mangola Kango. 2003. 'Normal Nutrition Fundamental Management'. RBSA Publishers, Jaipur.

Manoranjan K. 2002. Food Analysis and Quality. Kalyani Publishers, New Delhi.

McWilliams, M. and Paine, H. 1984. Modern food preservation. Surjeet Publications, Delhi. 198p.

Messina, M and Messina, V. 1996. Nutritional Implications of Dietary Phytochemicals. In: Dietary Phytochemicals in Cancer Prevention and Treatment. Plenum Press. New York, 401: 207-212.

Molins, R. A. 2001. Food Irradiation: Principles and Applications. New York: John Wiley and Sons. 488p.

Mudambi R, Sumathi and Rajagpal M V 1983), "Foods and Nutrition", Willey Eastern Ltd, Second Edition, New Delhi.

Mudambi, S R and Rajagopal, M V. 2008. "Food Science", New Age International Private Limited Publishers, New Delhi, 412p.

Mudambi, S. R.; Rao, S. M. and Rajagopal, M.V. 2006. Food Science. 2nd ed. New Age International (P) Limited, Publishers, New Delhi. 248p.

Mudambi, Sumathi V. and Rajagopal, M.V. 2001. Fundamentals of Foods and Nutrition. New Age International (P) Ltd. Publishers, New Delhi. 405p.

National Institute of Nutrition [NIN]. 2005. Dietary Guidelines for Indians - A Manual, Hyderabad.

Pandey, P. H. 2015. Principles and practices of post harvest technology. Kalyani Publishers, New Delhi. 539p.

Pariza M. 1999. Functional foods: technology, functionality, and health benefits. Nutrition, Today, 34: 150–151.

Parker, R. 2003. Introduction to food science. Delmar Thomson Learning, New York. 636p.

Peckham, G C, Jeanne, H and Freeland-Graves. 1979. Foundations of food preparation. MacMillan Publishing Co, Inc, New York, 616p.

Potter and Joseph. 2007. "Food Science" CBS Publishers and Distributors, Chennai, Indai, 602p.

Potter, N.N. and Hotchkiss, J.H. 1998. Food science. 5th Ed. Aspen Publishers, Maryland, USA. 608 p.

Raheena Beegum, M. 2011. 'Speaking of Child Care and Nutrition' Sterling Publishers, Karnataka, India, 128p.

Raheena Begum, M. 2008. 'A textbook of Foods, Nutrition and Dietetics'. Sterling Publishers private limited, Karnataka, India, 395p.

Rahman, M.S. 2007. Handbook of food preservation. 2nd Ed. CRC Press Inc., Boca Raton. 1068p.

Ramakrishnan S, Prassanan, K G and Rajan R. 1989. "Text book of Medical Bio chemistry". 2nd Edition, orient Longman limited.

Ranganna S. 1986. Manual of analysis of fruits and vegetable products, Mc. Graw Hill International Publishers, New Delhi.

Ranken M D. 2000. Handbook of meat product technology, Blackwell Science, United States, 212p.

Reddy, B S, Rao, C V. Rivenson, A and Kelloff, G 1993. Chemoprevention of colon carcinogenisis by organosulfur compounds. Cancer Res, 53: 3493–3498.

Robinson C H, Lawler, M. R, Chenoweth W and Garwick A. 1986. Normal and Therapeutic Nutrition, 17th edition, MacMillan International, Canada, 759p.

Roday, S. 1998. Food Hygiene and Sanitation. Tata McGraw-Hill Publishing Company limited, New Delhi. 315p.

Roday, S. 2012 "Food Science and Nutrition" Tata McGraw-Hill publishing company limited, New Delhi, 425p.

Seema Yadav, 1997. 'Basic Principles of Nutrition' Anmol Publicaitons Private Limited,

Shaifur Rehman. M. 2007. Handbook of food preservation. Second Edition. CRC, Press. London. 1088p.

Shakuntala Manay, N and Shadakshara Swamy M. 2001. Food- facts and principles. New Age International Limited Publishers, New Delhi, 520p.

Sharma, J L and Silvano Caralli. 1998. A dictionary of Food and Nutrition, CBS publishers and Distributors Private Limited, Chennai, 820p.

Shubangini A Joshi, (1998): Nutrition and Dietetics, Tata Mc Graw Hill Pub. Co. Ltd., New Delhi.

Shubhagini Joshi. 2010. Nutrition and Dietetics (with Indian Care Studies), Tata Mc Graw Hill Education Private Limited.

Shubhangini A and Joshi. 2010. 'Nutrition and Dietetics' Tata McGraw Hill, Pvt Ltd., NewDelhi.

Singh. N. P. 2007. Fruits and vegetables preservation. Oxford Book Company. Jaipur. India. 351p.

Sivasankar, B. 2002. Food processing and preservation. Prentice Hall of India Pvt. Ltd. 360p.

Smith, J. S. and Hui, Y.H. 2004. Food Processing: principles and applications. Blackwell Publishing, USA. 511p.

Srilakshmi B. 2003. 'Food Science Laboratory Manual'. Scitech Publications, Chennai.

Srilakshmi B. 2015. Food Science, 6th Edition, New Age International Publishers, New Delhi.

Srilakshmi B. 2016. 'Nutrition Science', 5th Edition, New Age International Publishers, New Delhi.

Srilakshmi, B. 2014. Dietetics, 7th edition, New Age International Publishers, New Delhi.

Srilakshmi, B. 2016. "Food Science", New Age International (P) Ltd, Publishers, New delhi.

Srivastava, R.P. and Kumar, S. 2006. Fruit and vegetable preservation. Principles and practices. International Book Distributing Co., Lucknow. 474p.

Steinmetz K A and Potter J D. 1991. Vegetables, fruit and cancer II: Mechanisms Cancer Causes and Control, 2: 427–442.

Sudheer, K.P. and Indra, V. 2007. Post-harvest technology of horticultural crops: Vol. 7. Horticulture science series. New India Publishing Agency, New Delhi. 290p.

Sumati Mudambi R and Shalini Rao. 1989. Food science, New Age International Publishers, New Delhi,

Swaminathan M S. 2009 Handbook of Food and Nutrition, BAPPCO.

Swaminathan M. 1979. "Food Science and Experimental Foods". Ganesh and Co, Madras.

Swaminathan M. 1988. Instructional manual for detecting Food adulterants- Home kit.

Swaminathan. 1995. "Food and Nutrition". Volume I, 2nd Edition, The Bangalore Printing and publishing Co Limited, Bangalore.

Swaminathan. M. 2008. 'Food and Nutrition Volume – 1'. The Bangalore Publishing Co. Limited, Bangalore.

Vijay Sethi and Shruti Sethi. 2006. Processing of Fruits and Vegetables for Value Addition. Indus Publishing. New Delhi.194p.

Whitney E L, Cataldo C B and Debruyne L K. 2015. Nutrition and Diet Therapy, 9th edition, Cengage Learning, U.S.A.

Index

V

W

X

Y

Z